AN INTRODUCTION TO GEOLOGICAL STRUCTURES AND MAPS

AN INTRODUCTION TO GEOLOGICAL STRUCTURES AND MAPS

Sixth Edition

G. M. Bennison
Formerly Senior Lecturer in Geology, University of Birmingham

and

K. A. Moseley
Head of Physics and Geology, Monmouth School

A member of the Hodder Headline Group
LONDON • NEW YORK • SYDNEY • AUCKLAND

First edition published in the United Kingdom 1964
Sixth edition published in Great Britain in 1997 by
Arnold, a member of the Hodder Headline Group
338 Euston Road, London NW1 3BH

Second impression 1998

Co-published in the US, Central and South America by
John Wiley & Sons, Inc.
605 Third Avenue, New York, NY 10158-0012

British Library Cataloguing in Publication Data
A catalogue record for this book is available from the British Library

Library of Congress Cataloging-in-Publication Data
Bennison, George Mills.
An introduction to geological structures and maps. – 6th ed. /
G.M. Bennison and K.A. Moseley.
p. cm.
Includes index
ISBN 0-470-23743-0
1. Geology, Structural–Maps. 2. Geological mapping.
I. Moseley, K. A. (Keith Anthony), 1956-. II. Title
QE601.2.B46 1997 96-29760
551.8′022′3–dc21 CIP

ISBN 0 340 69240 5
ISBN 0 470 23743 0 (USA)

Publisher: Laura McKelvie
Production Editor: James Rabson
Production Controller: Rose James
Cover designer: Stefan Brazzo

Composition by Scribe Design, Gillingham, Kent
Printed by The Bath Press, Bath, UK.

CONTENTS

PREFACE TO THE CURRENT EDITION

The proposal that there should be a new (sixth) edition of this well-established small volume provided the welcome opportunity to introduce new ideas and to add new chapters, as well as to rewrite two chapters and eliminate minor ambiguities from two existing maps. It also provided the opportunity to seek a co-author with fresh ideas and a close working relationship with 'A'-level and Open University students of Geology. I was pleased when Keith Moseley accepted my invitation to share the work of this new material with me. It was decided that the need was not for a more advanced book in this field – there are now a number of these – but to permit more graduated steps to the advanced problems which this book includes.

I should like to express my particular thanks to Dr R. Pickering for his constructive criticism, both of my prose and my maps, and for his diligence and willingness to pass on his original ideas.

G.M. Bennison
Colwall Green, Malvern, 1997

PREFACE TO PREVIOUS EDITIONS

This book is designed primarily for university and college students taking geology as an honours course or as a subsidiary subject. Its aim is to lead the student by easy stages from the simplest ideas on geological structures right through the first year course on geological mapping, and much that it contains should be of use to students of geology at GCE 'A' level. The approach is designed to help the student working with little or no supervision: each new topic is simply explained and illustrated by figures, and exercises are set on succeeding problem maps. If students are unable to complete the problems they should read on to obtain more specific instructions on how the theory may be used to solve the problem in question. Problems relating to certain published geological survey maps are given at the end of most chapters. Some of the early maps in the book are of necessity somewhat 'artificial' so that new structures can be introduced one at a time thus retaining clarity and simplicity. Structure contours (see p. 7) are seldom strictly parallel in nature; it is therefore preferable to draw them freehand, though – of course – as straight and parallel as the map permits. In all cases except the 'three-point' problems, the student should examine the maps and attempt to deduce the geological structures from the disposition of the outcrops in relation to the topography, as far as this is possible, before commencing to draw structure contours.

The author wishes to thank Dr F. Moseley for making many valuable suggestions when reading the manuscript of this book, Dr R. Pickering and Dr A.E. Wright for their continuing help and interest.

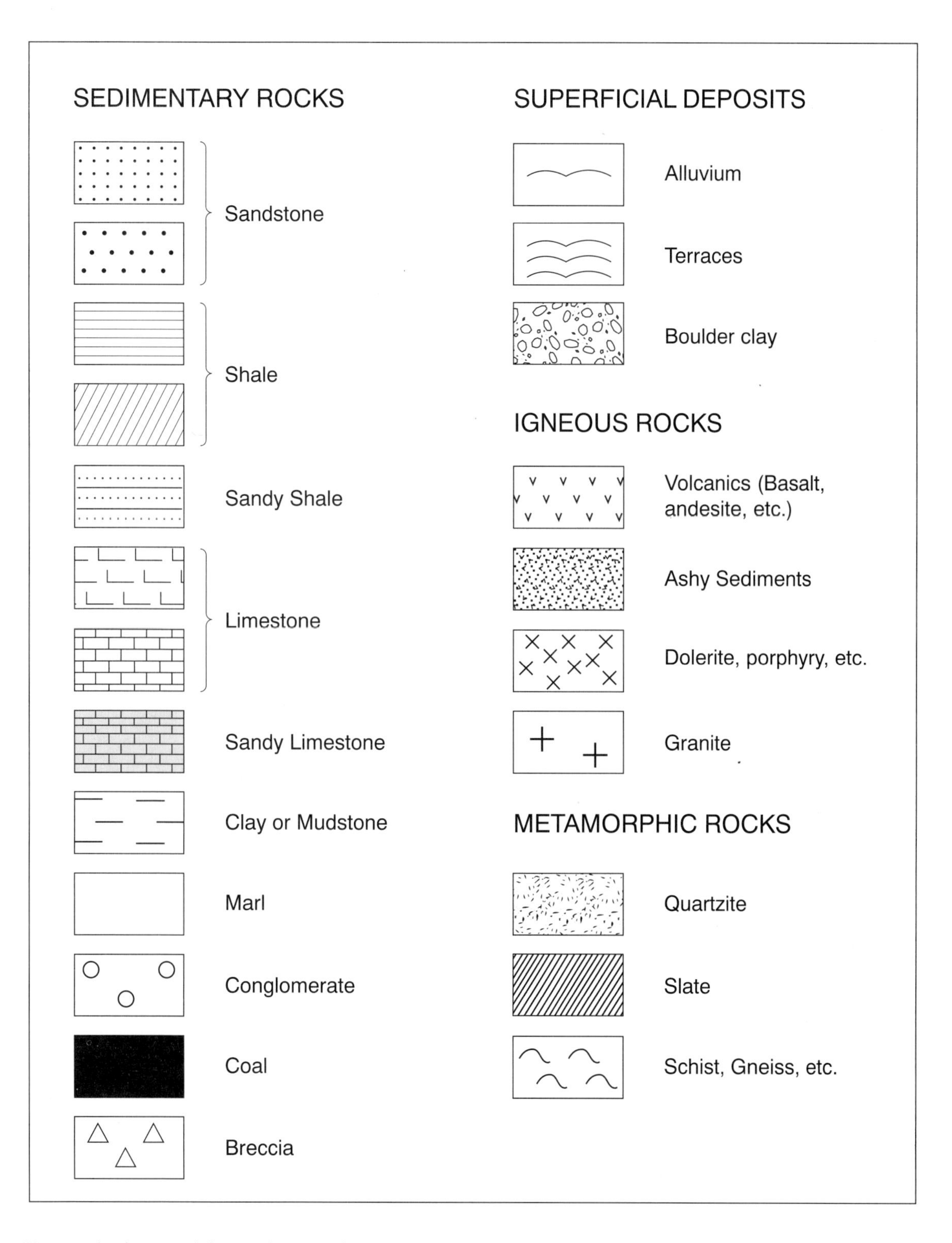

Key to shading widely used on geological maps and text figures

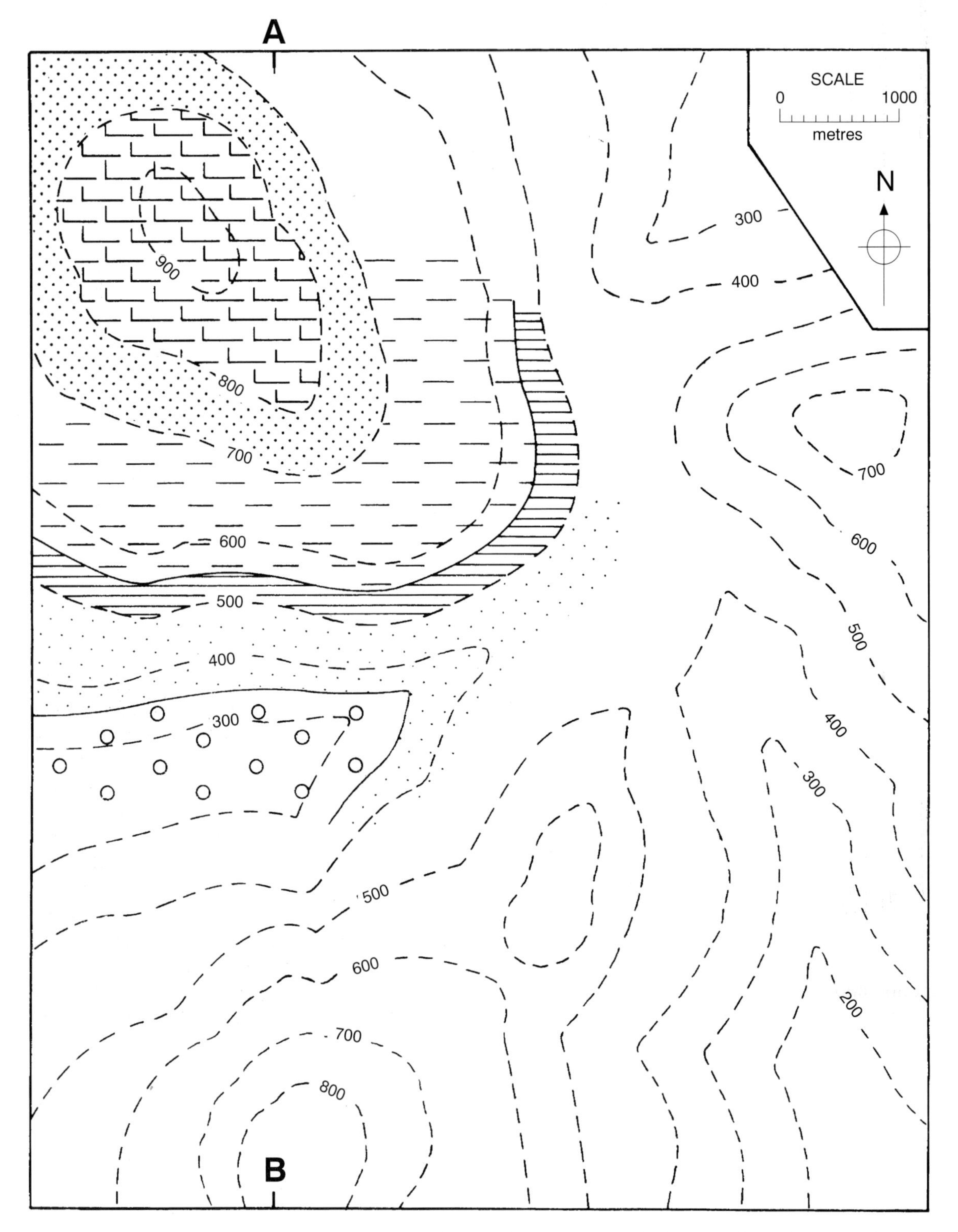

Map 1 The geological outcrops are shown in the north-west corner of the map. It can be seen that the beds are horizontal as the geological boundaries coincide with, or are parallel to, the ground contour lines. Complete the geological outcrops over the whole map. Indicate the position of a spring-line on the map. How thick is each bed? Draw a vertical column showing each bed to scale: 1 cm = 100 m. Draw a section along the line A–B. (Contours in metres.)

1

HORIZONTAL AND DIPPING STRATA

CONTOURS

Hills and valleys are usually carved out of layered sequences of rock, or strata, the individual members – or beds – differing in thickness and in resistance to erosion. Hence diverse topography (surface features) and landforms are produced. Only in exceptional circumstances is the topography eroded out of a single rock-type.

In the simplest case we can consider strata as horizontal. Rarely are they so in nature; they are frequently found elevated hundreds of metres above their position of deposition, and tilting and warping has usually accompanied such uplift. The pattern of outcrops of the beds where the strata are horizontal is a function of the topography; the highest beds in the sequence (the youngest) will outcrop on the highest ground and the lowest beds in the sequence (the oldest) will outcrop in the deepest valleys. Geological boundaries will be parallel to the contour lines shown on a topographic map for they are themselves contour lines, since a contour is a line joining an infinite number of points of the same height.

Section drawing

Draw a base line the exact length of the line A–B on Map 1 (19.0 cm). Mark off on the baseline the points at which the contour lines cross the line of section: for example, 8.5 mm from A mark a point corresponding to the intersection of the 700 m contour. From the baseline erect a perpendicular corresponding in length to the height of the ground and, since it is important to make vertical and horizontal scales equal wherever practicable, a perpendicular of length 14 mm must be erected to correspond to the 700 m contour (since 1000 m = 2 cm and 100 m = 2 mm) (Fig. 1). Sections can readily be drawn on metric squared paper (or on 1/10" in some cases).

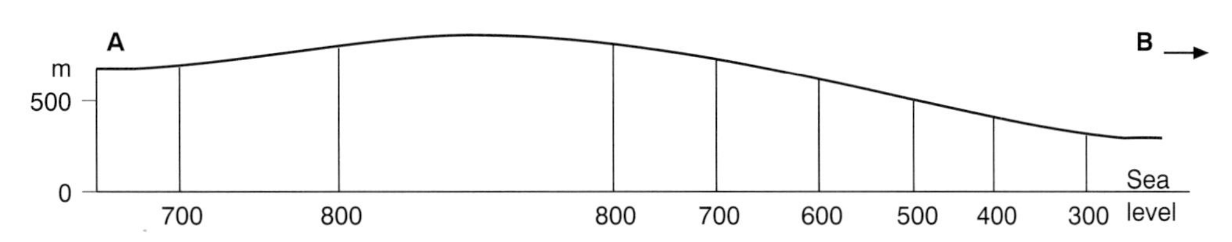

FIGURE 1 Part of a section along the line A–B on Map 1 to show the method of drawing the ground surface (or profile).

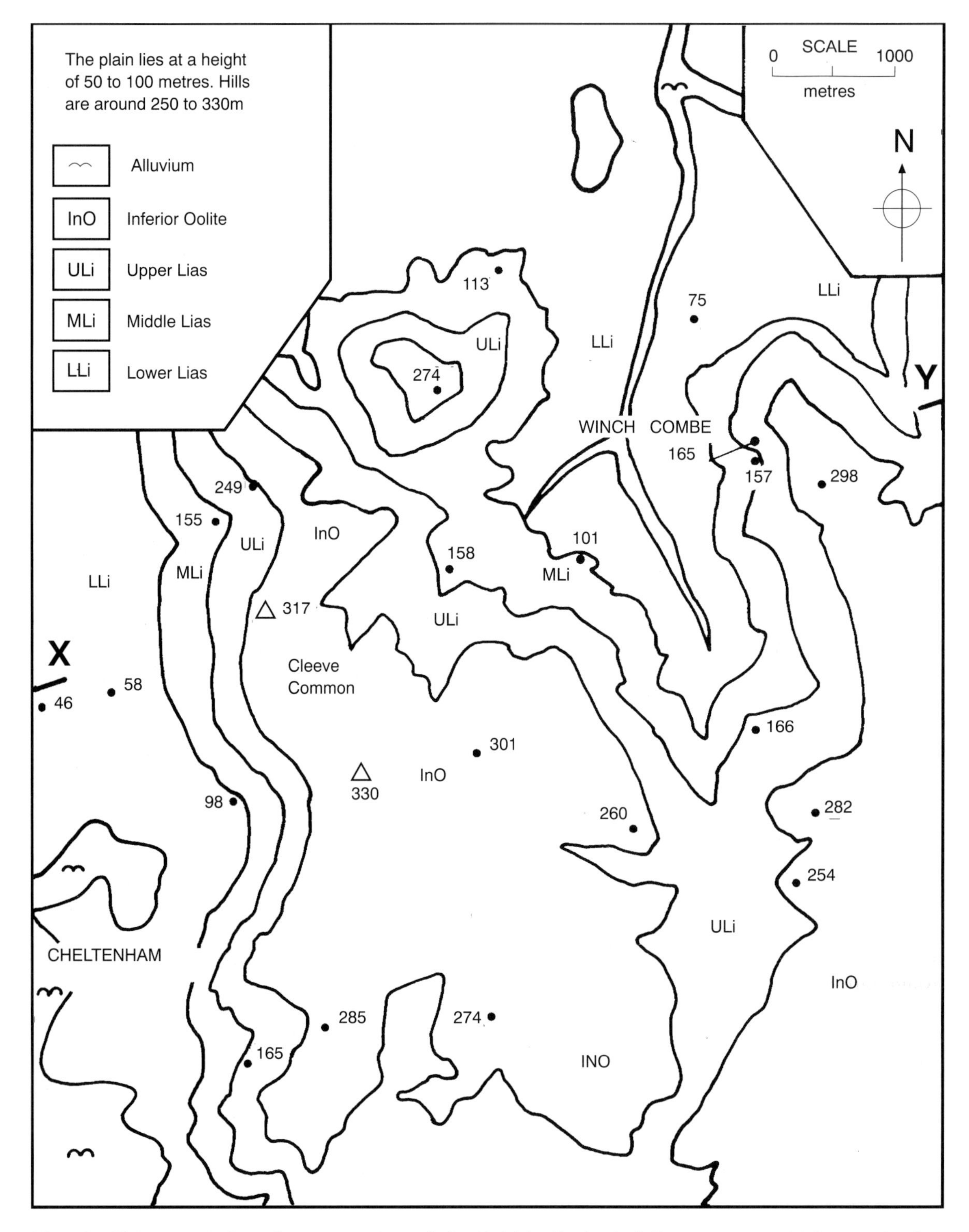

Map 2 This map is based on a portion of the British Geological Survey map of Moreton-in-the-Marsh, 1:50,000 scale, Sheet 217. (It is slightly simplified and reduced to fit the page size.) Reproduced by permission of the Director, British Geological Survey: NERC copyright reserved. Draw a section along the line X–Y to illustrate the geology, using a vertical scale of 1 cm = 200 m.

Map 2 shows an area of the Cotswold Hills and adjacent lowlands where the strata are virtually horizontal. Geological boundaries, therefore, are parallel to topographic contours, a point made on page 1. Contours have been omitted for clarity but the general heights of hills and plain are given. These, together with the altitude of a number of points (spot-heights and triangulation points) enable us to draw a sufficiently accurate topographic profile of a section across the map.

Vertical Exaggeration

The horizontal scale of a map has been determined at the time of the original survey. Commonly, we shall find it given as 1:50,000 – 1 cm on the map representing 50,000 cms or 500 metres on level ground (1:50,000 is the 'Representative Fraction'). Maps on the scale of 1:25,000 and 1:10,000 are common. On older maps one inch represented one mile (1:63,360) and on USA maps two miles to the inch is a usual scale.

If a similar scale to the horizontal is used vertically we shall often find that the section is difficult to draw and that it is very difficult to include the geological details. For example, on Map 2 the highest hill on the section at 317 m would be only just over 0.6 cms above the sea-level base line. A suggested vertical scale of 1 cm = 200 m, with horizontal scale predetermined as 1 cm = 500 m, gives a vertical exaggeration of 500/200 or 5/2 or 2.5.

Complications arise where the strata are inclined and further considerations of vertical exaggeration are discussed on p. 12.

When you have drawn the geological section on the profile provided, turn to page 115 where the geological section has been drawn on a true scale (horizontal and vertical scales the same, Fig. 46a). It has also been redrawn with a vertical exaggeration of ×5 (horizontal scale 1 cm = 500 m, vertical scale 1 cm = 100 m). The true scale section illustrates the practical difficulty, while the latter section, Fig. 46b, shows that too great a vertical exaggeration produces an unnatural distortion (and in the case of dipping strata causes other problems dealt with at the end of this chapter). Where vertical exaggeration is necessary from a practical point of view, it should be kept to a reasonable minimum.

Dip

Inclined strata are said to be dipping. The angle of dip is the maximum angle measured between the strata and the horizontal (regardless of the slope of the ground) (Fig. 2).

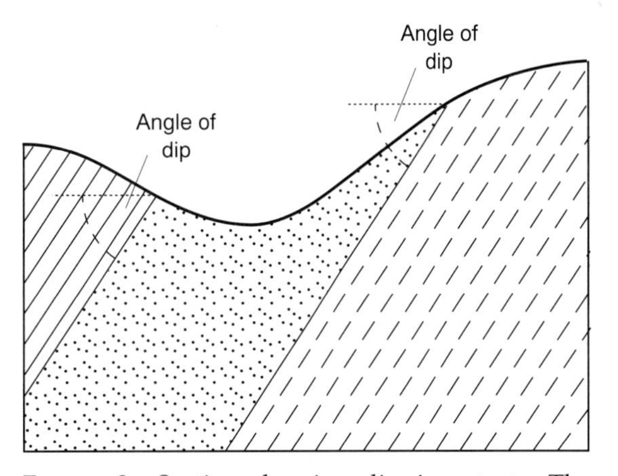

Figure 2 Section showing dipping strata. The angle of dip is measured from the horizontal.

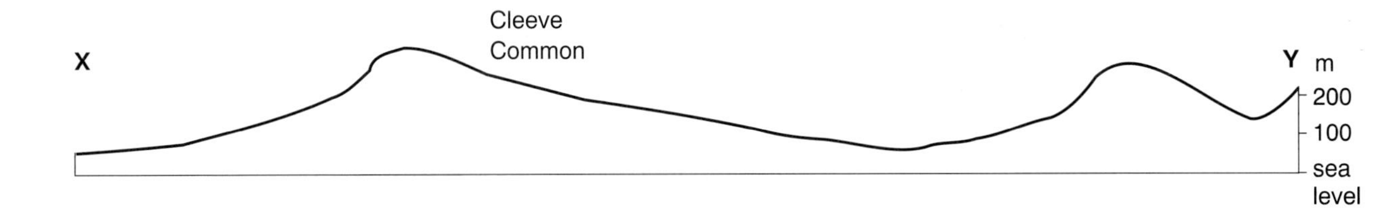

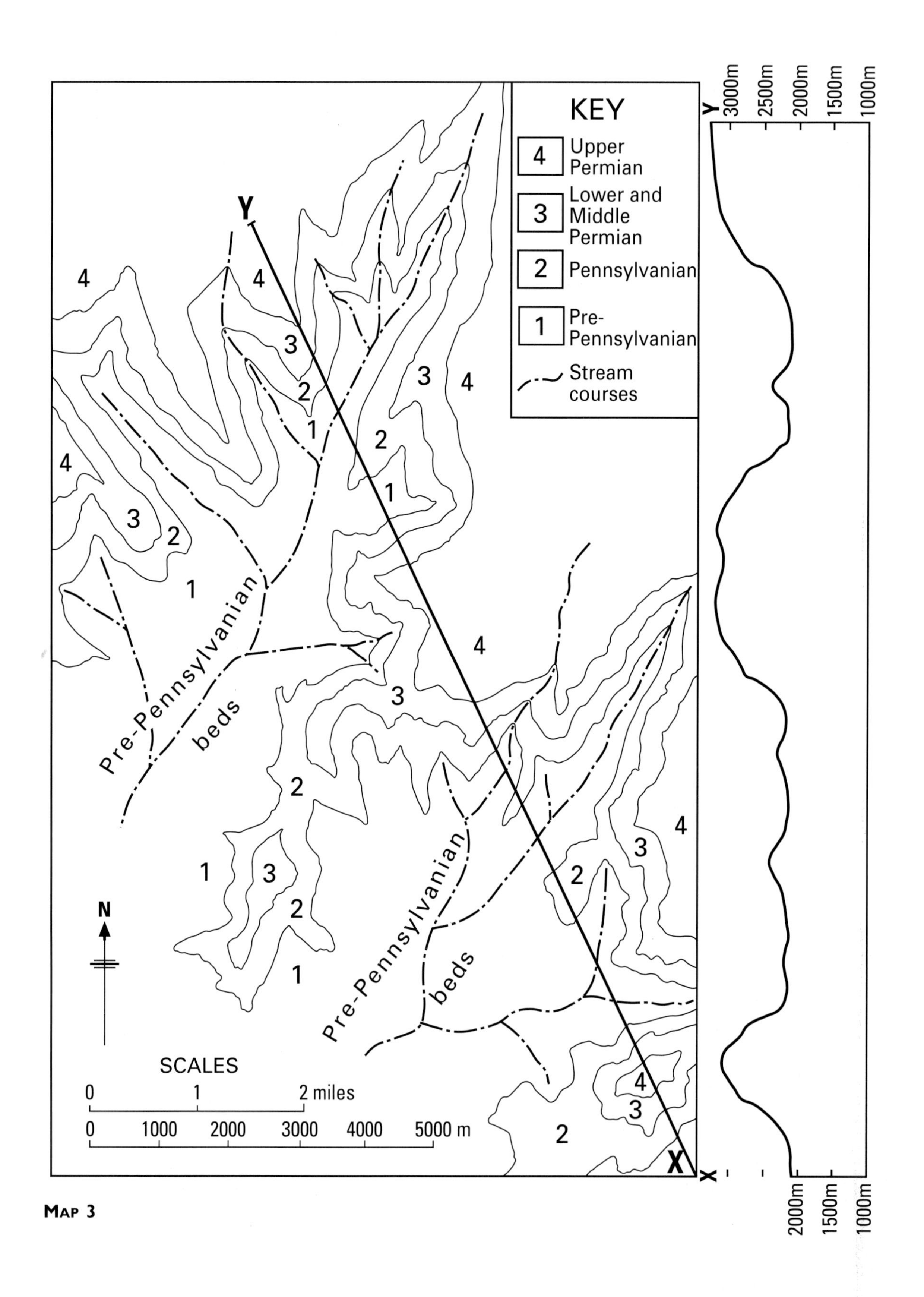
KEY
4 Upper Permian
3 Lower and Middle Permian
2 Pennsylvanian
1 Pre-Pennsylvanian
Stream courses
Y
X
Pre-Pennsylvanian beds
Pre-Pennsylvanian beds
N
SCALES
0 1 2 miles
0 1000 2000 3000 4000 5000 m
3000m
2500m
2000m
1500m
1000m
2000m
1500m
1000m

Map 3

The direction of dip is given as a compass bearing reading from 0 to 360°. For example, a typical reading would be 12/270. The first figure is the angle of dip, the angle that strata make with the horizontal. The second figure is the direction of that dip measured round from north in a clockwise direction (in this example due west). In North America, where the use of the Brunton Compass is almost universal, dip directions are more commonly given in relation to the main points of the compass, e.g. 10° west of south (i.e. 190°).

In a direction at right angles to the dip the strata are horizontal. This direction is called the strike (Fig. 3). An analogy may be made with the lid of a desk. A marble would roll down the desk lid in the direction of maximum dip. The edge of the desk lid, which is the same height above the floor along the whole of its length, i.e. it is horizontal, is the direction of strike.

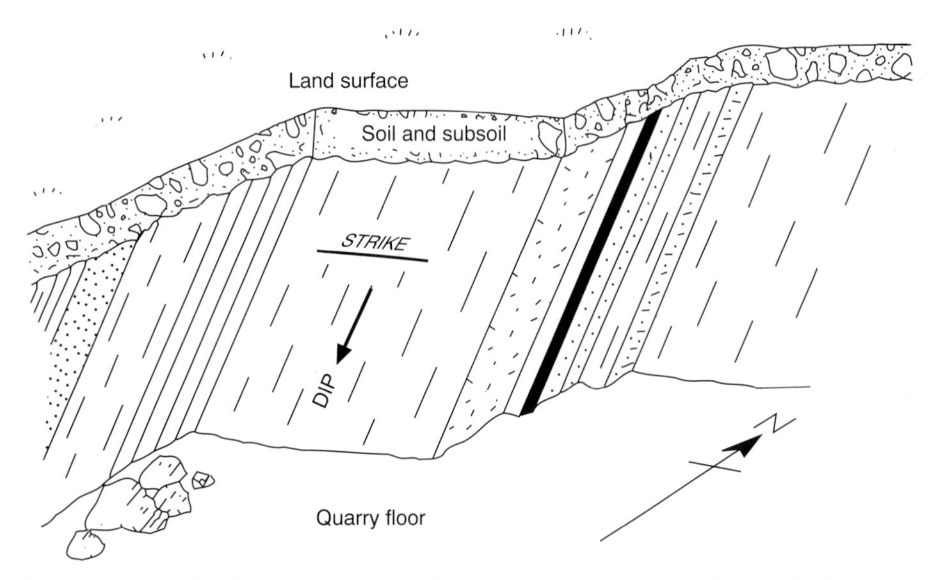

FIGURE 3 Southerly dipping strata in a quarry. Note the relationship between the directions of dip and strike.

MAP 3 This shows a simple geological map based on the geology of the Bright Angel area of the Grand Canyon, Arizona. The strata are horizontal. How can horizontal strata give rise to such a complicated pattern of outcrops? It is because the topography is complicated. A plateau has been very deeply dissected by a dendritic pattern of rivers and tributaries. The high altitude of this region and low base level give the streams great erosive power which in this fairly arid region of sparse vegetation has resulted in steep-sided canyons with many branches. You may find it helpful to colour the different strata to make the outcrop pattern clearer. For simplicity, contours have been omitted from the map. Of course, the outcrops of the geological boundaries between strata of different ages are themselves contours, since the strata are horizontal.

The base of the Pennsylvanian is at a height of 2250 m and it is 500 m thick. The Lower and Middle Permian together are 250 m thick. Insert the geology on the profile provided along the line of section X–Y.

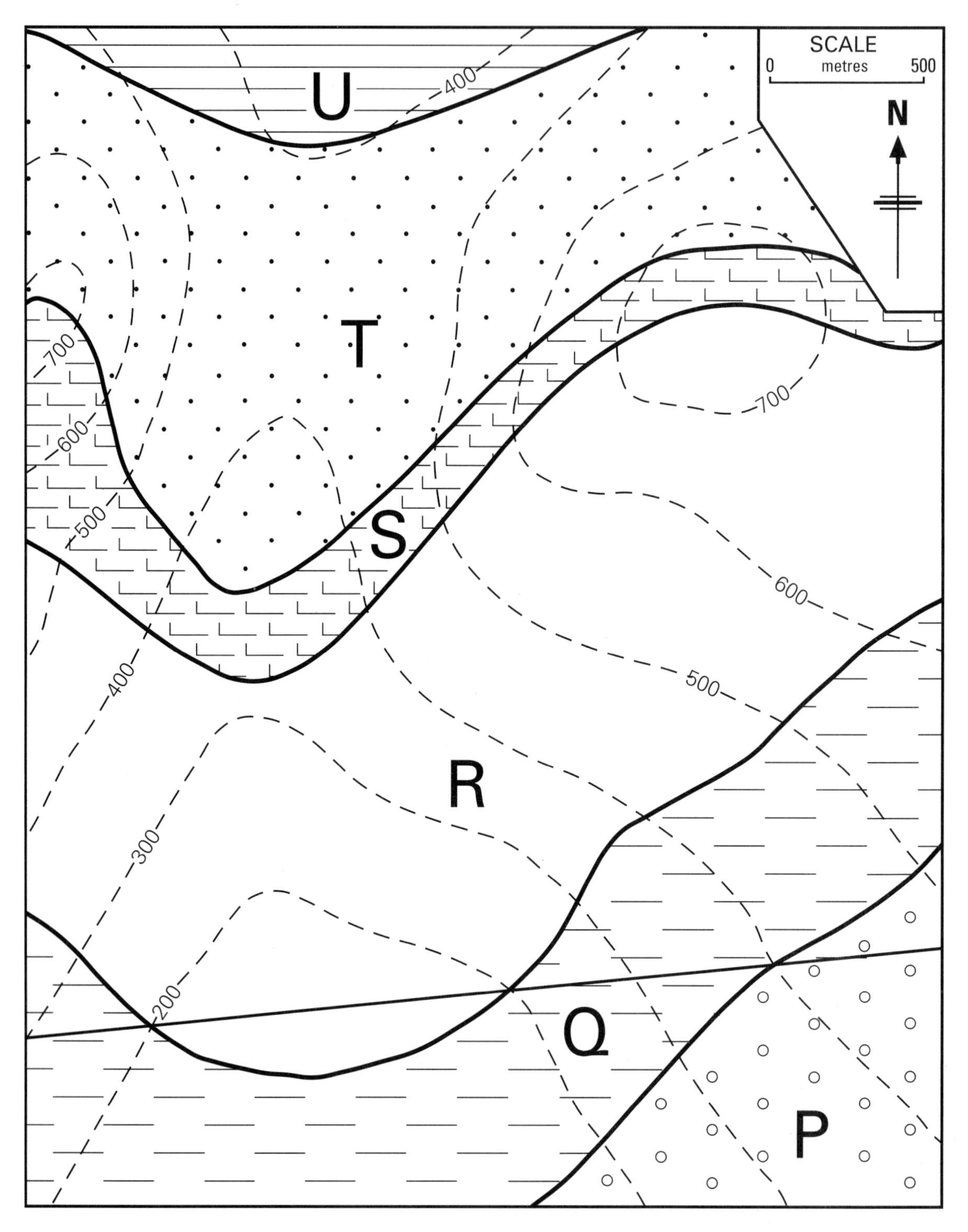
SCALE
0 metres 500
N
U
T
S
R
Q
P
400
700
600
500
400
300
200
700
600
500

Map 4

STRUCTURE CONTOURS (= STRIKE LINES)

Just as it is possible to define the topography of the ground by means of contour lines, so we can draw contour lines on a bedding plane. These we call structure contours or strike lines, the former since they join points of equal height, the latter since they are parallel to the direction of strike. The terms are synonymous, but for the purposes of this book the term 'structure contour' will be used.

CONSTRUCTION OF STRUCTURE CONTOURS

The height of a geological boundary is known where it crosses a topographic contour line. For example, the boundary between beds S and T on map 4 cuts the 700 m contour at three points. These points lie on the 700 m structure contour which can be drawn through them. Since these early maps portray simply inclined plane surfaces, structure contours will be straight, parallel and – if dips are constant – equally spaced.

Having found the direction of strike to be 85°, we know that the direction of dip is at right angles to this, but we must ascertain whether the dip is 'northerly' or 'southerly'. A second structure contour can be drawn on the same geological boundary S–T through the two points where it cuts the 600 m contour. From the spacing of the structure contours we can calculate the dip or gradient of the beds (Fig. 4(a)):

Gradient	=	700 m – 600 m in 1.25 cm
i.e.	=	100 m in 1.25 cm.

As the scale of the map is given as 2.5 cm = 500 m, 100 m in 1.25 cm = 100 in 250 m. Hence, **the gradient is 1 in 2.5, to 175°**

Frequently it is more convenient to utilize gradients, although on geological maps the dip is always given as an angle. By simple trigonometry we see that the angle of dip in the above case is that angle which has a tangent of 1/2.5 or 0.4, i.e. 22°, to 175° (Fig. 4(b)).

Section drawing

The topographic profile is drawn by the method already described on page 1. The geological boundaries (interfaces) can be inserted in an analogous way by marking the points at which the line of section is cut by

MAP 4 The continuous lines are the geological boundaries separating the outcrops of the dipping strata, beds P, Q, R, S, T and U. Examine the map and note that the geological boundaries are not parallel to the contour lines but, in fact, intersect them. This shows that the beds are dipping. Before constructing structure contours can we deduce the direction of dip of the beds from the fact that their outcrops 'V' down the valley? Can we deduce the direction of dip if we are informed that Bed U is the oldest and Bed P is the youngest bed of the sequence? Draw structure contours for each geological interface[1] and calculate the direction and amount of dip. (Contours in metres.) Instructions for drawing structure contours are given below and one structure contour on the Q/R geological boundary has been inserted on the map as an example.

[1]Some confusion may arise since the term geological boundary is often applied both to the interface (or surface) between two beds and also to the outcrop of that interface. It seems a satisfactory term to employ, however, since the two are related and the context generally avoids ambiguity.

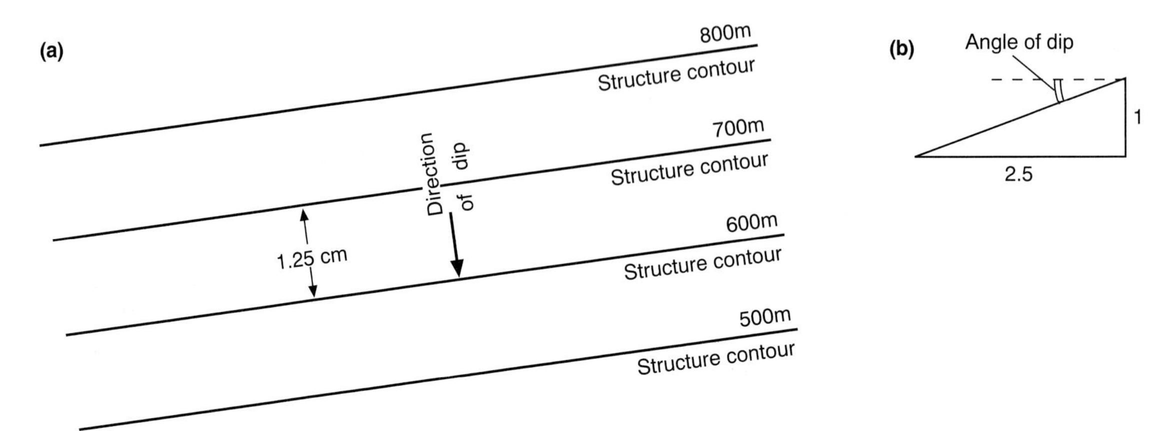

FIGURE 4 (a) Plan showing structure contours and (b) section through contours showing the relationship between dip and gradient.

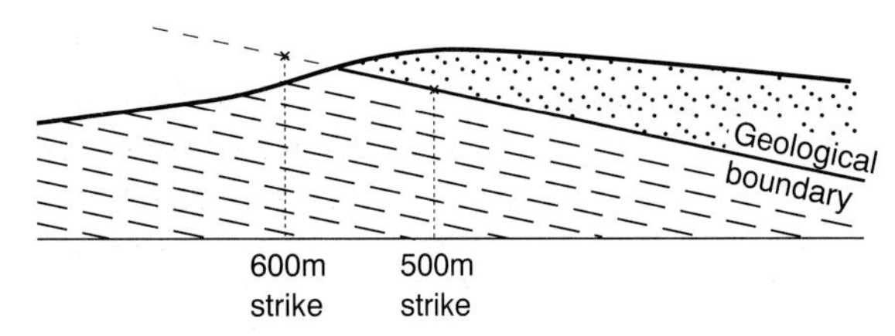

FIGURE 5 Section to show the method of accurately inserting geological boundaries.

structure contours. Perpendiculars are then drawn from the base line, of length corresponding to the height of the structure contours (Fig. 5).

TRUE AND APPARENT DIP

If the slope of a desk lid, or of a geological boundary (interface) or bedding plane, is measured in any direction between the strike direction and the direction of maximum dip, the angle of dip in that direction is known as an apparent dip (Fig. 6(a)). Its value will lie between 0° and the value of the maximum or true dip. Naturally occurring or man-made sections through geological strata (cliffs, quarry faces, road and rail cuttings) are unlikely to be parallel to the direction of true dip of the strata. What may be observed in these sections, therefore, is the dip of the strata in the direction of the section, i.e. an apparent dip (somewhat less than the true dip in angle). The trigonometrical relationship is not simple:

$$\text{Tangent apparent dip} = \text{tangent true dip} \times \cos\beta$$

(see Fig. 6(b)).

However, the problem of apparent dip calculation is much simplified by considering it as a gradient. Just as the gradient of the bed in the direction of maximum dip is given by the spacing of the structure contours (1.9 cm = 380 m in Fig. 6(b), representing a gradient of 1 in 3.8, since the scale of the map is 1 cm = 200 m), so the gradient in the direction in which we wish to obtain the apparent dip is given by the structure contour spacing measured in that direction (3.25 cm = 650 m in Fig. 6(b), representing a gradient of 1 in 6.5).

In road and rail cuttings the direction of dip of strata is of vital importance to the stability of the slopes. Where practicable a cutting would be parallel to the direction of dip of the strata, minimizing slippage into the cutting since there would be no component of dip at right angles to the face of the cutting. Factors other than

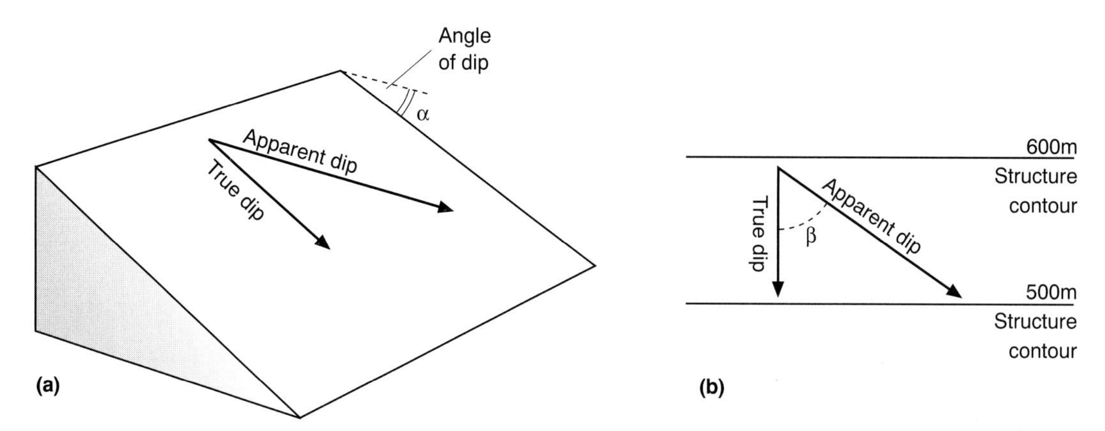

FIGURE 6 (a) Diagram and (b) plan or map of structure contours to illustrate the relationship between true and apparent dip.

geological ones determine the siting and direction of cuttings. Frequently they are not parallel to the dip of the strata and, as a result, we see an apparent dip in the cutting sides.

In the event of a geological section being at right angles to the direction of dip it will be, of course, in the direction of strike of the beds. There will be no component of dip seen in this section and the beds will appear to be horizontal. (Of course, close examination of such a quarry face or cutting will reveal that the strata are not horizontal but are dipping towards or away from the observer.)

CALCULATION OF THE THICKNESS OF A BED

On Map 5 it can be seen that the 1100 m structure contour for the geological boundary D–E coincides with the 1000 m structure contour for boundary C–D. Thus, along this strike direction, the top of bed D is 100 m higher than its base. It has a vertical thickness of 100 m. This is the thickness of the bed that would be penetrated by a borehole drilled at point X.

Turn back to Map 4. The 200-m structure contour for the Q–R boundary, which has already been inserted on the map, passes through the point where the P–Q boundary is at 400 m. Since structure contours are all parallel – where beds are simply dipping as in this map – this line is also the 400-m structure contour for the P–Q boundary. It follows that bed Q has a vertical thickness of 200 m. You will find that on many problem maps bed thickness is often 100 m, 200 m or some multiple of 50 m in order to produce a simpler problem.

VERTICAL THICKNESS AND TRUE THICKNESS

Since the beds are inclined, the vertical thickness penetrated by a borehole is greater than the true thickness measured perpendicular to the geological boundaries (interfaces) (Fig. 7). The angle α between VT (vertical thickness) and T (true thickness) is equal to the angle of dip.

$$\text{Now cosine } \alpha = \frac{T}{VT}$$

$$\therefore \mathbf{T = VT \times cosine\ \alpha}$$

The true thickness of a bed is equal to the vertical thickness multiplied by the cosine of the

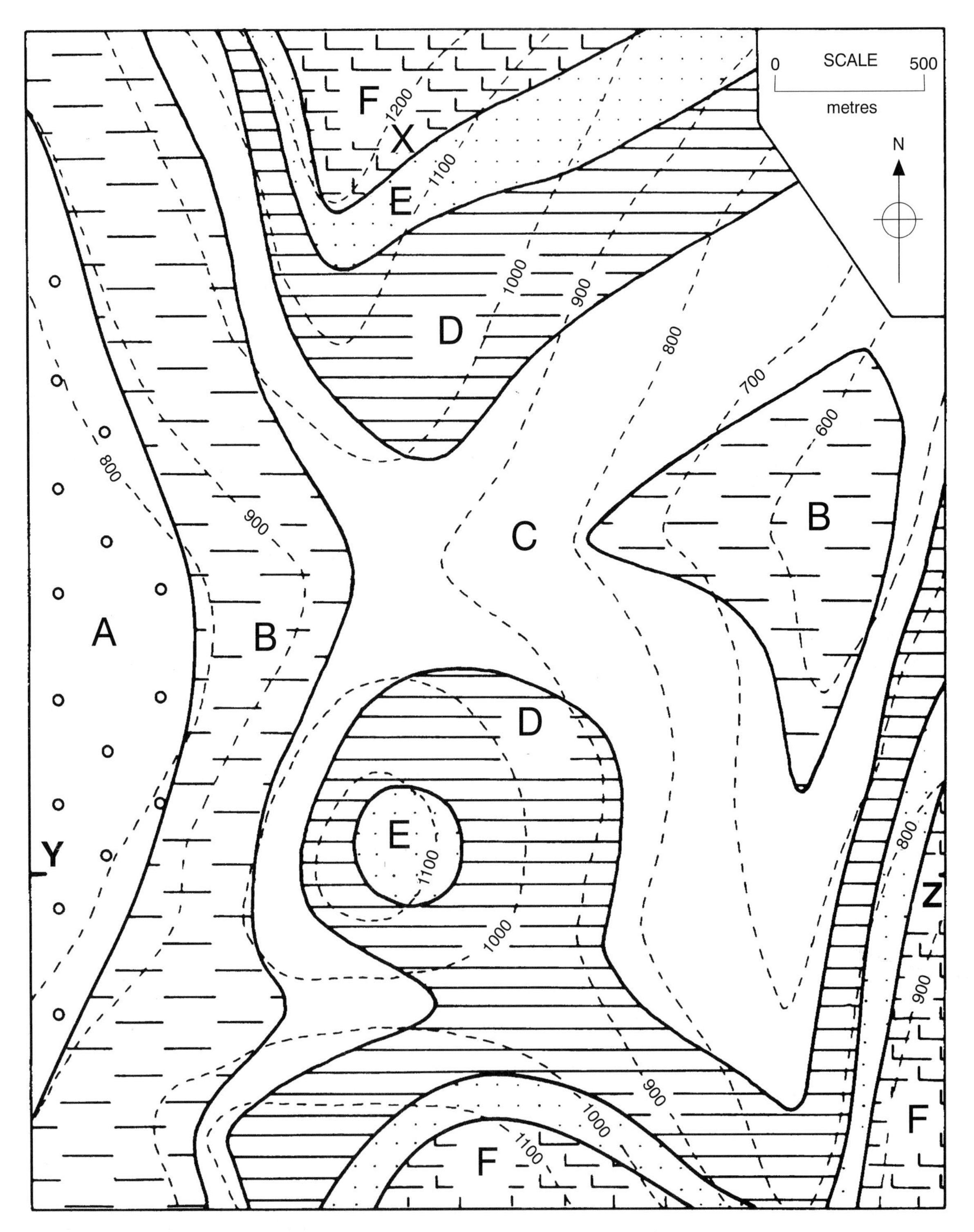

MAP 5 Draw structure contours on the geological boundaries. Give the gradient of the beds (dip). Draw a section along the east–west line Y–Z. Calculate the thicknesses of beds B, C, D and E. Indicate on the map an inlier and an outlier.

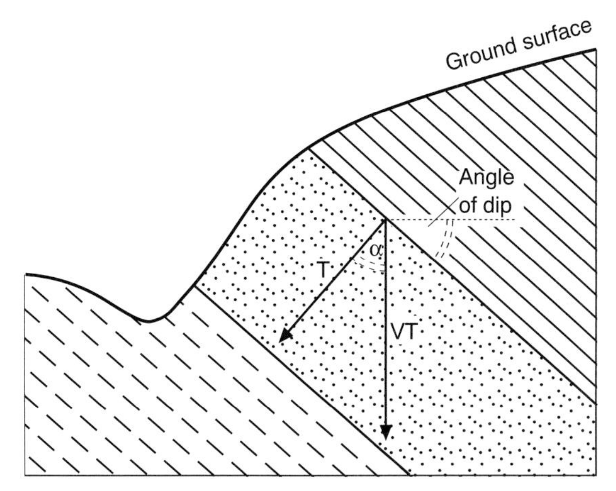

FIGURE 7 Section showing the relationship between the vertical thickness (VT) and the true thickness (T) of a dipping bed.

angle of dip. Where the dip is low (less than 5°) the cosine is high (over 0.99) and true and vertical thicknesses are approximately the same (see Table 2, p. 117).

WIDTH OF OUTCROP

If the ground surface is level, the width of outcrop of a bed of constant thickness is a measure of the dip (Fig. 8(a)).

Naturally, where beds with the same dip crop out on ground of identical slope the width of outcrop is related directly to the thickness of the beds (Fig. 8(c)), Table 3, p. 117.

More generally, beds crop out (or outcrop) on sloping ground and width of outcrop is a function of dip and slope of the ground as well as bed thickness. In Fig. 8(b) beds Y and Z have the same thickness (and dip) but, due to the different angles of intersection with the slope of the ground their widths of outcrop (W) are very different. ($T_1 = T_2$ but W_1 is much greater than W_2.)

It will be noted that in the case of horizontal strata the geological boundaries are parallel to the topographic contours. In dipping strata the geological boundaries cross the topographic contours, and with irregular topography the

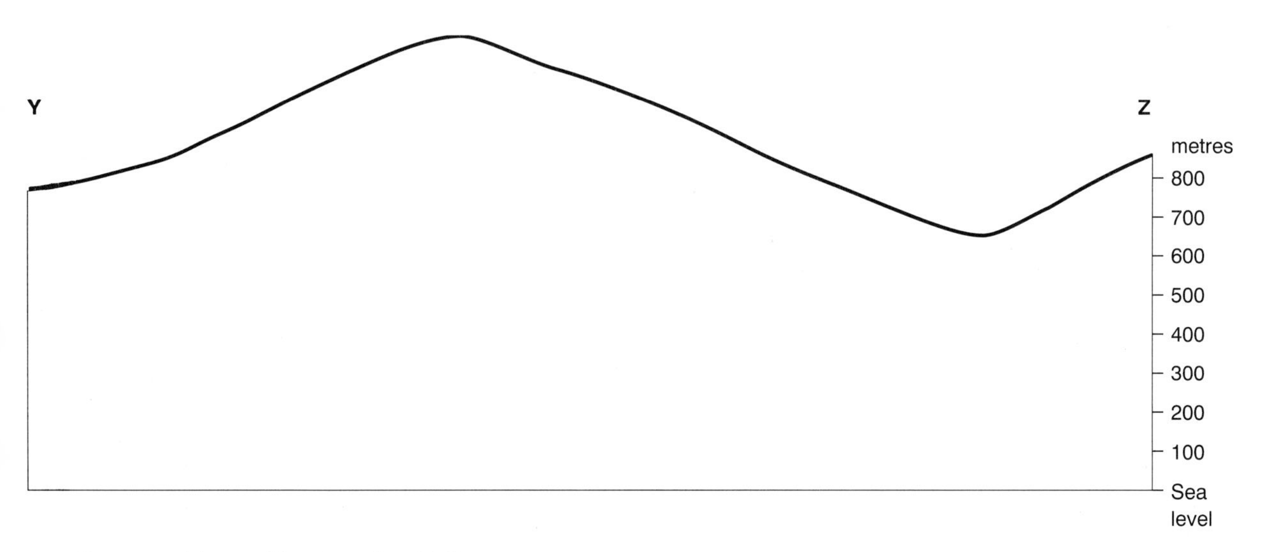

Topographic profile of section Y–Z.

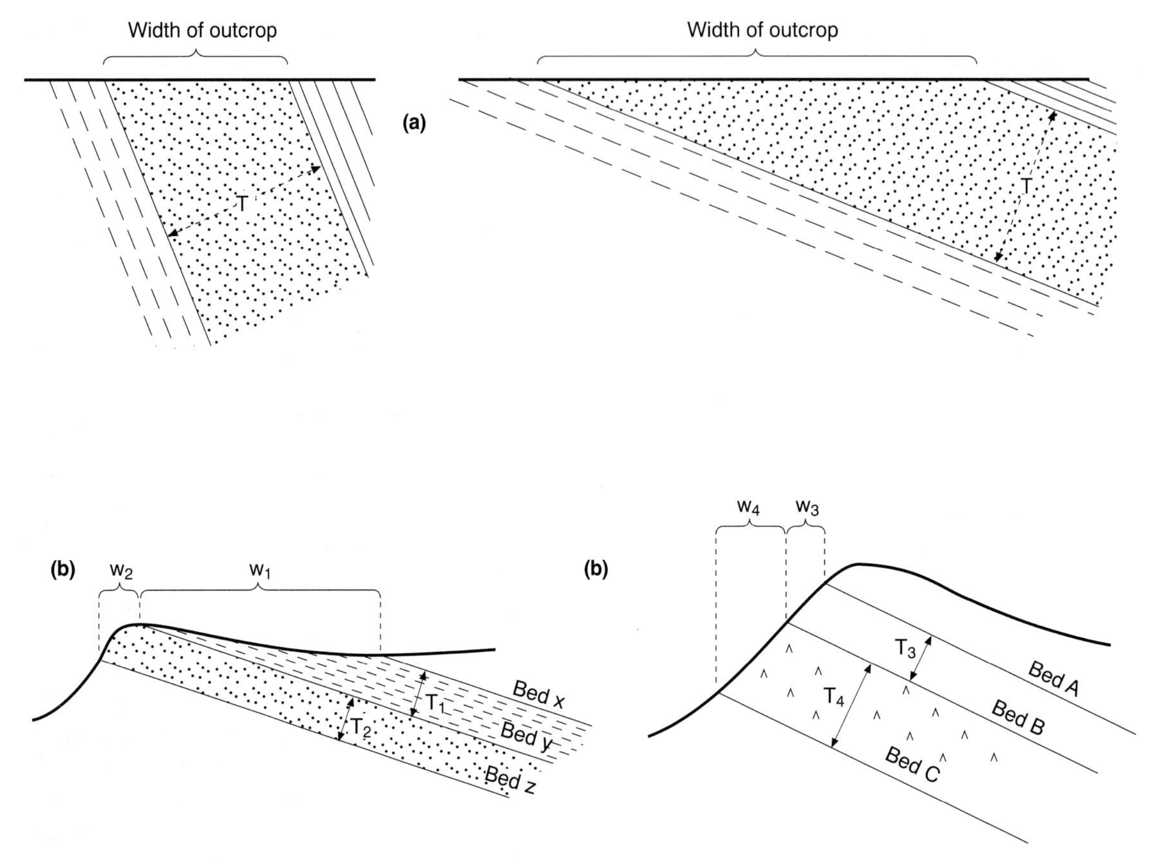

FIGURE 8 (a) Sections showing the different widths of outcrop produced by a bed of the same thickness (T) with high dip and low dip. (b) Beds of the same thickness ($T_1 = T_2$) outcropping on differing slope. (c) Beds of different thickness ($T_4 = T_3 \times 2$) outcropping on a uniform slope.

steeper the dip the straighter the outcrops. In the limiting case, that of vertical strata, outcrops are straight and unrelated to the topography.

INLIERS AND OUTLIERS

An outcrop of a bed entirely surrounded by outcrops of younger beds is called an inlier. An outcrop of a bed entirely surrounded by older beds (and so separated from the main outcrop) is called an outlier. In Map 5 these features are the product of erosion on structurally simple strata and are called 'erosional' inliers and outliers. (See p. 47 for details of other inliers and outliers.)

Vertical Exaggeration

It is very difficult to make the vertical scale equal to the horizontal scale for a one-inch to the mile map, which is of course, 1″ = 5280 feet. It is necessary to introduce a vertical exaggeration which should be kept to the minimum practicable. No hard and fast rule can be made. The British Geological Survey (BGS) commonly employs a vertical exaggeration (VE) of two or three, but sections provided on BGS maps range from true scale to a vertical exaggeration of as much as ×10 where this is necessary. The immensely useful 'Geological Highway Map' series of the USA (published by and obtainable from the AAPG, Tulsa, Oklahoma) illustrate the effects of an overlarge VE of ×20. With one-inch to the mile maps, a vertical scale of 1″ = 1000 feet is particularly convenient and the VE at approximately ×5¼ is generally acceptable.

It is essential to realize that where the section is drawn with a vertical exaggeration of three, for example, the *tangent* of the angle of dip must be multiplied by three to find the angle of dip appropriate to this section. (It is *not the angle of dip* which is multiplied by three.) A dip of 20° should be shown on a section with a VE of ×3 as a dip of 48°. Look up the tangents and check that this is correct.

EXERCISES USING GEOLOGICAL SURVEY MAPS

1. **Henley-on-Thames: 1:50,000 (sheet 254) Solid & Drift Edition** Examine the map and section, noting the relationship of topography to geology. (The Chalk is one of the most resistant formations in South East England, forming the high ground of the Chilterns, Downs, etc.) Note numerous outliers to the west of the main cuesta.

   A structure contour map is included. You will see that the structure contours are approximately parallel but not straight as on most problem maps which cover a smaller area as a rule. What do you conclude from the curvature of the structure contours?
2. **Aylesbury: 1″ (Sheet 238)** Excluding the rather extensive Pleistocene and Recent deposits (which form a quite thin superficial cover), the oldest strata are to be found in the north-west of the area with successively younger beds to the south-east. This gives the direction of dip. The amount of dip can best be determined by ensuring that beds are made the appropriate thickness on the section, e.g. Lower Chalk and Middle Chalk should each measure approximately 200 feet if the section has been correctly drawn. Draw a section along a line in a north-west–south-east direction across the map to illustrate the structure of the area.

Note on the Aylesbury and Henley-on-Thames sheets. When dealing with an area the size of one of these maps, it is found that strata are not uniformly dipping (as is the case on the simpler problem maps) but may be slightly flexured. Structure contours — if they could be drawn — would not be quite straight nor precisely parallel. The beds cannot be inserted on a section by constructing structure contours: they must be 'fitted' to the outcrop widths and drawn at their correct thicknesses, using information given in the stratigraphic column in the margin of the map.

Notes on BGS Maps The British Geological Survey produces a much wider variety of maps than formerly, both in range of scales employed and in the kind of information included.

Most of England and parts of Wales, together with about half of Scotland has now been covered by maps published on the 1:50,000 scale. Some areas are covered by still available and most useful one-inch to the mile sheets. Both solid geology editions and drift editions have been produced for some areas. For the purposes of investigating geological structures the solid editions are the more useful. However, drift editions, showing the superficial deposits, are of vital importance to the engineering geologist concerned with planning motorways, dams and foundations. In areas where drift deposits are not extensive a single edition combining Solid and Drift is usually published.

2

'THREE-POINT' PROBLEMS

If the height of a bed is known at three or more points (not in a straight line), it is possible to find the direction of strike and to calculate the dip of the bed, provided dip is uniform. This principle has many applications to mining, opencast and borehole problems encountered by applied geologists and engineers but this chapter deals only with the fundamental principle and includes a few simple problem maps.

The height of a bed may be known at points where it outcrops or its height may be calculated from its known depth in boreholes or mine shafts. If the height is known at three points (or more) only one possible solution exists as to the direction and amount of dip, and this can be simply calculated.

CONSTRUCTION OF STRUCTURE CONTOURS

Note: since Map 6 portrays a coal seam, and an average seam is of the order of 2 m or less in thickness, on the scale of 2.5 cm = 500 m its thickness is such that it can satisfactorily be represented by a single line on the map. There is no need to attempt to draw structure contours for the top and the base of the seam – on this scale they are essentially identical.

Observe the height of the seam at points A, B and C where it outcrops. Join with a straight line the highest point on the coal seam, C (600 m) to the lowest point on the seam, A (200 m). Divide the line A–C into four equal parts (since 600 m – 200 m = 400 m). As the slope of the seam is constant we can find a point on AC where the seam is at a height of 400 m (the mid-point). We also know that the seam is at a height of 400 m at point B. A straight line drawn through these two points is the 400 m structure contour. On a simply dipping stratum such as this, all structure contours are parallel. Construct the 200 m structure contour through point A, and the 300 m, the 500 m and the 600 m structure contours – the latter through point C. Having now established both the direction and the spacing of the structure contours, complete the pattern over the whole of the map.

DEPTH IN BOREHOLES

Relative to sea level, the height of the ground at the site of a borehole can be estimated from its proximity to contour lines and that of the coal seam at the same point on the map can be calculated from the structure contours. Quite simply, the difference in height between the ground surface and the seam is the depth to which the borehole must be drilled to reach the seam.

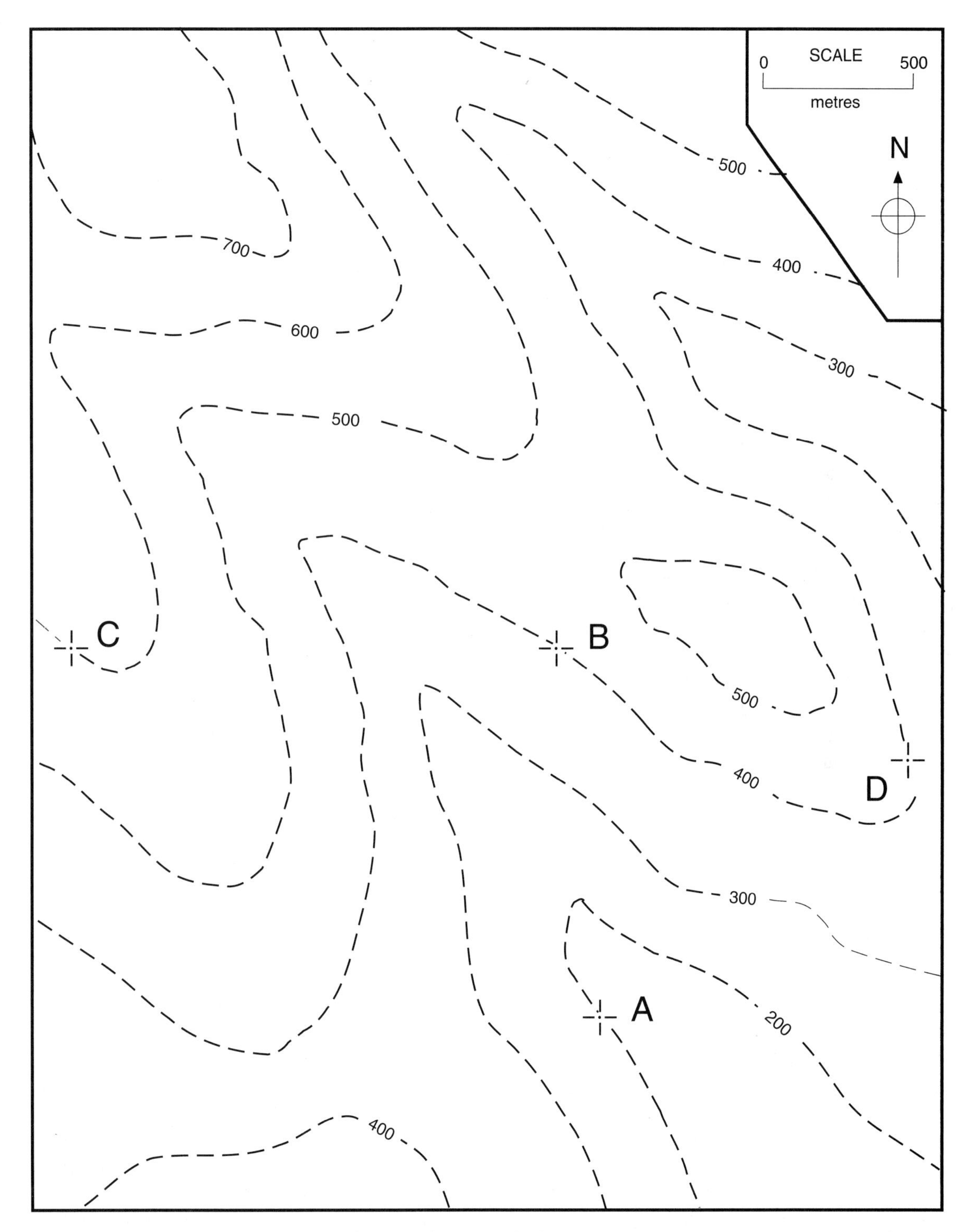

MAP 6 Deduce the dip and strike of the coal seam which is seen to outcrop at points A, B and C. At what depth would the seam be encountered in a borehole sunk at point D? Complete the outcrops of the seam. Would a seam 200 m below this one also outcrop within the area of the map? Contours in metres.

Notes on Map 6. At what depth would the coal seam be encountered in a borehole situated at point D? Since the borehole at D is sited on the 400 m topographic contour and (as you will see when you have constructed the structure contours for the coal seam) the 200 m structure contour passes through point D, the depth of the coal is equal to the height of the ground minus the height of the coal (400 m – 200 m). The depth to the coal is therefore 200 m.

Wherever a topographic contour intersects a structure contour we know both the height of the ground and the height of the coal bed. By simple arithmetic we can find the depth to the coal. If we join all the points where the seam is 200 m below ground level (as is the case in Borehole D) we are constructing a line called the 200 m isopachyte. (An isopachyte is defined on p. 47 where more complicated examples are to be found). On Map 6 draw in the isopachytes for 100 m and 200 m. They will tend to be roughly parallel to the outcrop (which is in effect the 'zero isopachyte'). If you find this problem difficult, return to it after you have worked out Map 16.

INSERTION OF OUTCROPS

The structure contours were drawn by ascertaining the height of the coal seam where it outcropped on contour lines. Wherever the seam – defined by its structure contours – is at the same height as the ground surface – defined by topographic contour lines – it will outcrop. We can find on the map a number of intersections at which structure contours and topographic contours are of the same height: the outcrop of the seam must pass through all these points.

Further, these points cannot be joined by straight lines. We must bear in mind that where the seam lies between two structure contours, e.g. the 300 m and 400 m, it can only outcrop where the ground is also at a height of between 300 m and 400 m, i.e. between the 300 m and 400 m topographic contours (Fig. 9). The outcrop of a geological boundary surface cannot, on a map, cross a structure contour or a topographic contour line except where they intersect at the same height.

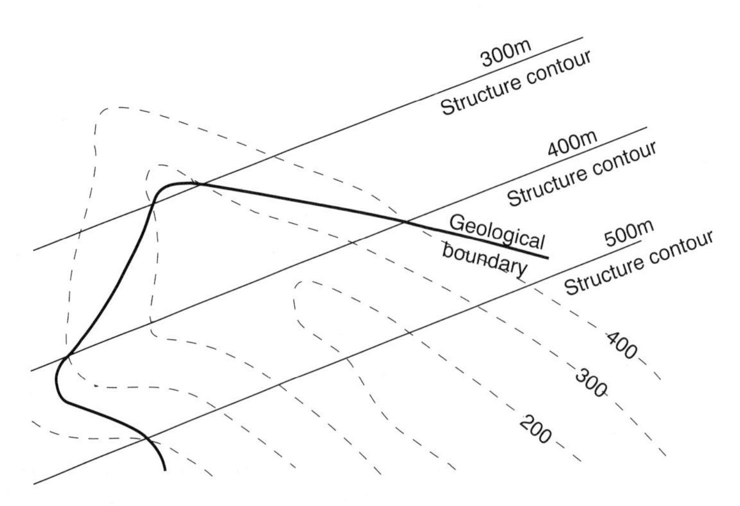

FIGURE 9 The insertion of a geological boundary on a map with topographic contours and structure contours.

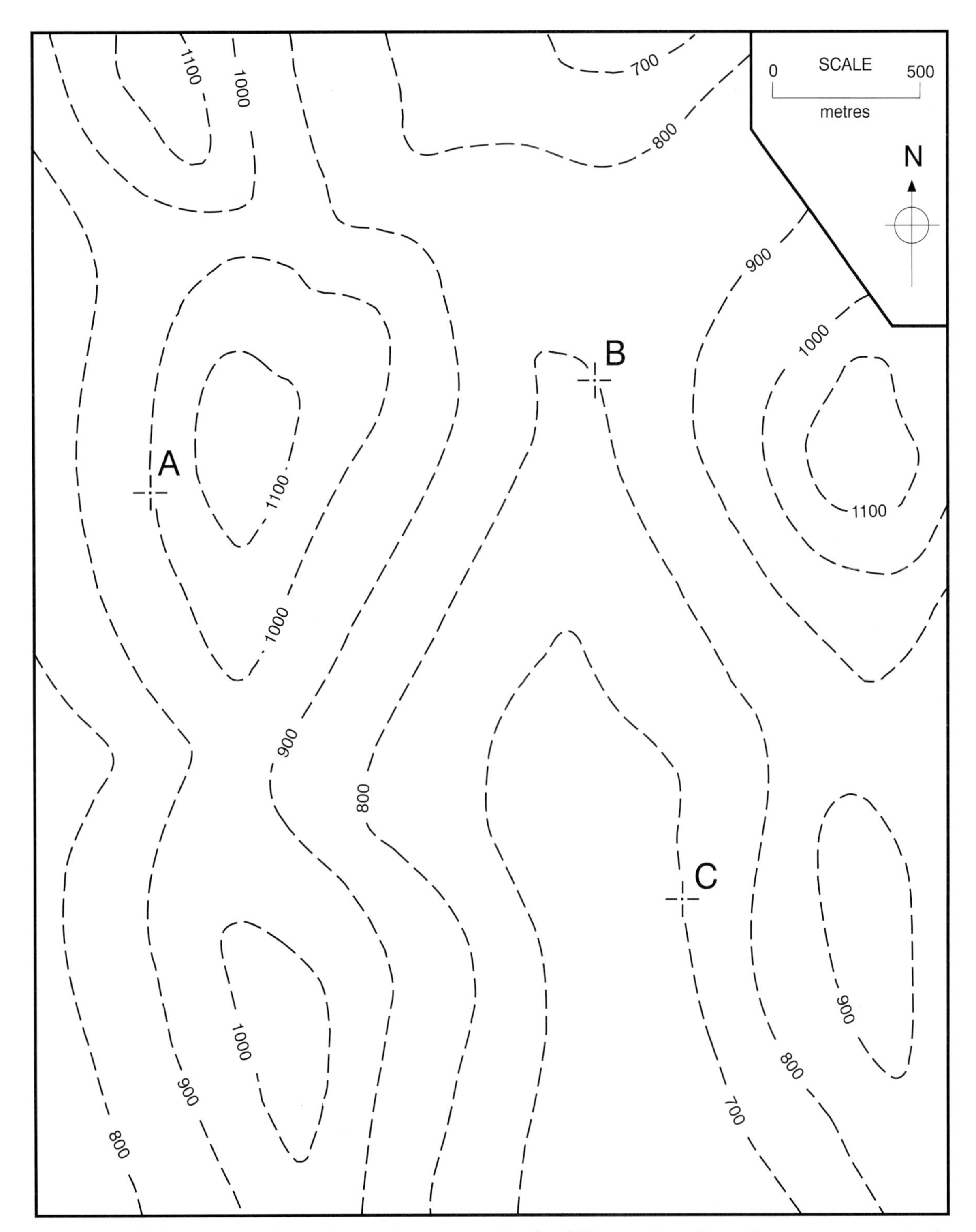

MAP 7 Borehole A passes through a coal seam at a depth of 50 m and reaches a lower seam at a depth of 450 m. Boreholes B and C reach the lower seam at depths of 150 m and 250 m, respectively. Having determined the dip and strike, map in the outcrops of the two seams (assume that the seams have a constant vertical separation of 400 m). Indicate the areas where the upper seam is at a depth of less than 50 m below the ground surface. It is necessary to first calculate the height (relative to sea level) of the lower coal seam at each of the points A, B and C where boreholes are sited.

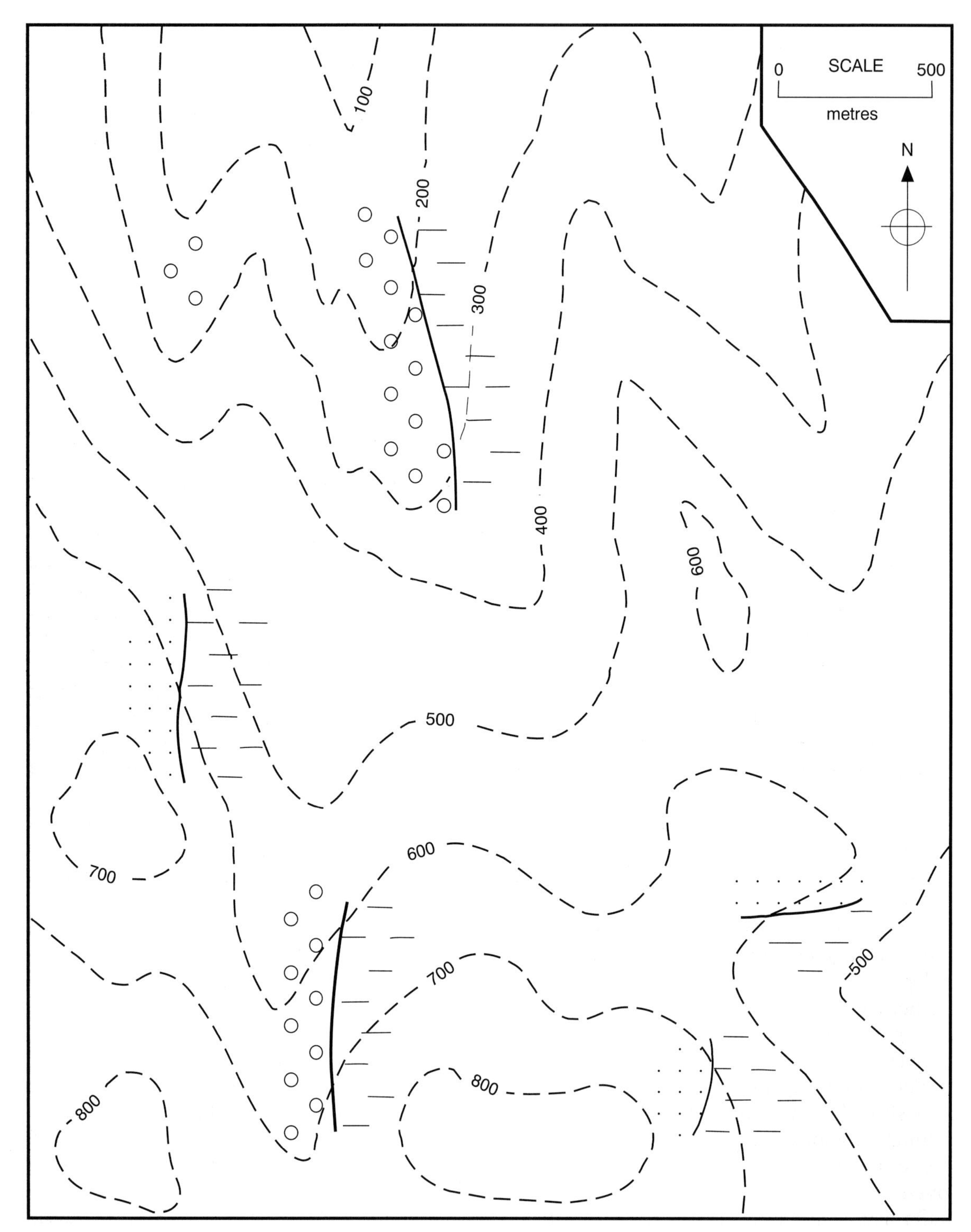

MAP 8 Three beds outcrop – conglomerate, sandstone and shale. Complete the geological boundaries between these beds, assuming that the beds all have the same dip. Indicate on the map an inlier and an outlier.

3

UNCONFORMITIES

In terms of geological history, an unconformity represents a period of time during which strata are not laid down. During this period, strata already formed may be uplifted and tilted by earth movements which also terminate sedimentation. The uplifted strata, coming under the effects of sub-aerial weathering and erosion, are 'worn down' to a greater or lesser extent before subsidence causes the renewal of sedimentation and the formation of further strata. As a result we find, in the field, one set of strata resting on the eroded surface of an older set of beds.

Many of the features that characterize unconformities cannot be deduced from map evidence alone. For example, only in the field can one observe the actual erosion surface, the presence of palaeosols, hardgrounds, etc. Derived fragments of the older strata may be redeposited in the post-unconformity strata (sometimes as a basal conglomerate) and the analogous but much rarer phenomenon of derived fossils may be seen. However, the evidence of a gap in the stratigraphic succession should be indicated in the stratigraphic column provided on a map and some of the above features may be referred to in the data given.

On a map the main evidence of unconformity is a difference in the dip and strike directions in the pre- and post-unconformity strata. Earth movements which, by uplift, terminated sedimentation and subsequent sinking which permitted resumed sedimentation frequently resulted in differences in dip and strike.

An exception to this can be found on the margins of the London Basin, where the Lower Tertiary beds rest unconformably on the Chalk with little difference in strike or dip yet, by comparison with successions of strata on the other side of the Channel, we know that this is a major unconformity representing a long period of time since, in Britain, the uppermost stage of the Chalk and the lowest two stages of the Tertiary are absent.

In some cases an unconformity represents a time interval of such length that the older strata were intruded by igneous rocks or were subjected to metamorphism.

OVERSTEP

Usually the lowest bed of the younger series of strata, having a quite different dip and strike from that of the older strata, rests on beds of different age. This feature (Fig. 10) is called overstep; bed X is said to overstep beds A, B, C, etc.

If the older strata were tilted before erosion took place, they meet the plane of unconformity at an angle, and there is said to be an 'angular unconformity' (Fig. 10).

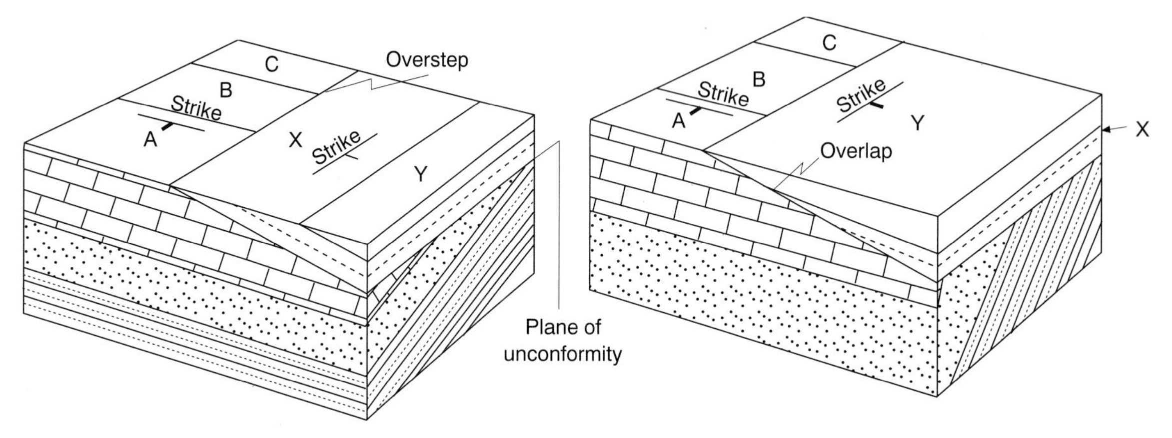

FIGURE 10 Block diagram of an angular unconformity.

FIGURE 11 Block diagram of an unconformity with overlap.

OVERLAP

As subsidence continues and the sea, for example, spreads further on to the old land area, successive beds are laid down, and they may be of greater geographical extent, so that a particular bed spreads beyond, or overlaps, the preceding bed. This feature (Fig. 11) may accompany an unconformity with or without overstep. Bed Y overlaps Bed X. (The converse effect, that of successive beds being laid down over a progressively contracting area of deposition, due to gradual uplift, is known as off-lap. Such a feature is rarely deducible from a geological map and will be discussed no further.)

Two principal types of unconformity are recognizable:

1. planes of marine erosion, readily definable by structure contours;
2. buried landscapes. Sub-aerial erosion did not produce a peneplain and the post-unconformity sediments were deposited on a very irregular surface of hills and valleys, gradually burying these features.

Problem maps 9, 10 and 17 are exercises based on type (1). Usually, it is necessary to look at a broader regional geology to see the effects of buried landscape. See below the note on the Assynt one-inch geological sheet.

SUB-UNCONFORMITY OUTCROPS

An unconformity represents a period of erosion. Tens or hundreds of metres of strata may be removed. How can we examine this phenomenon and deduce what remains of the older strata since they are now covered by post-unconformity strata? We could, of course,

MAP 9 Indicate on the map the outcrop of the plane of unconformity. Work out the dip and strike of the series of beds A to E and of beds P and Q. Note the difference in the strike direction of the two series, the most significant indication of unconformity from map evidence. Draw a section along the line X–Y on the profile provided.

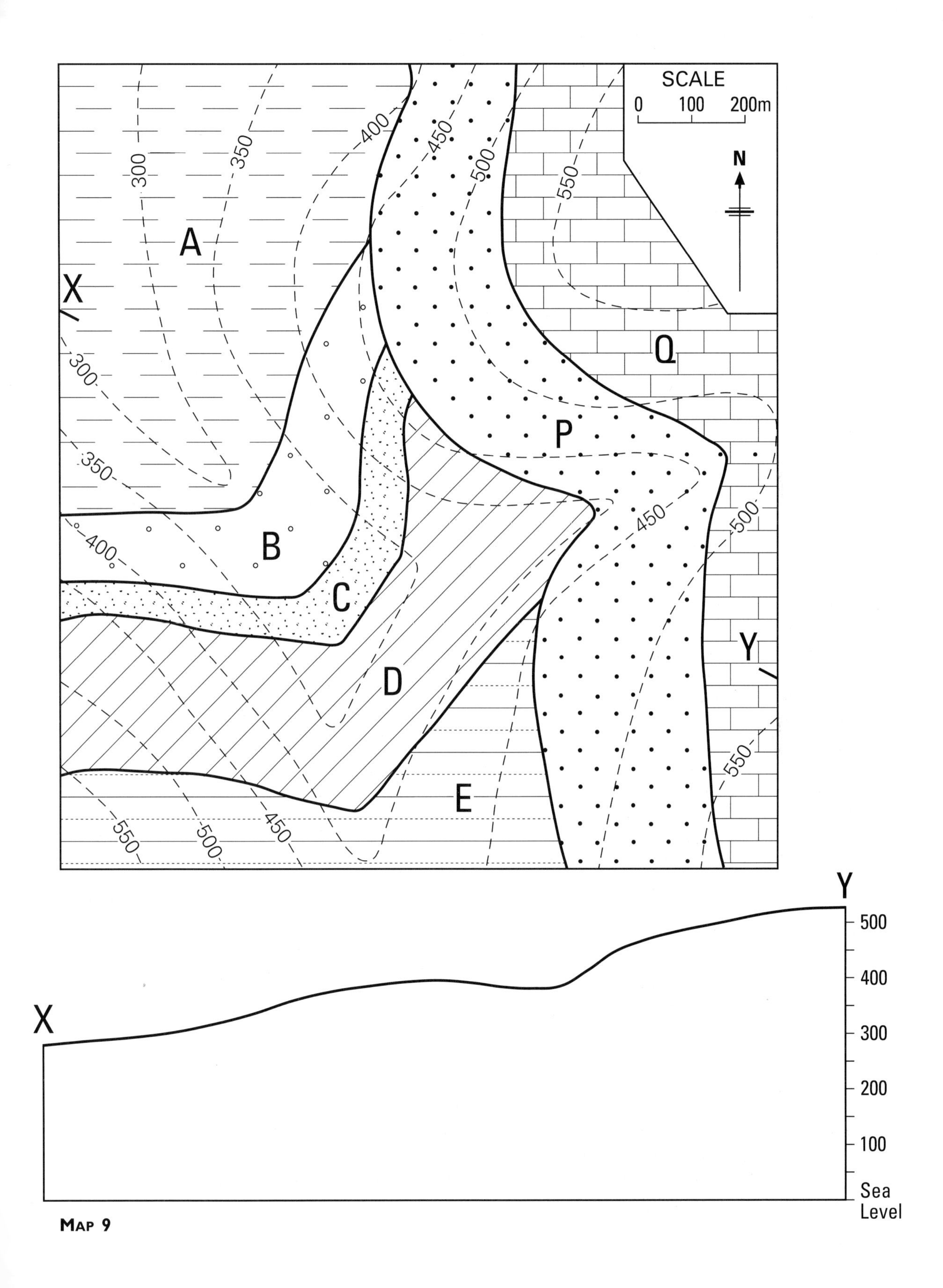

Map 9

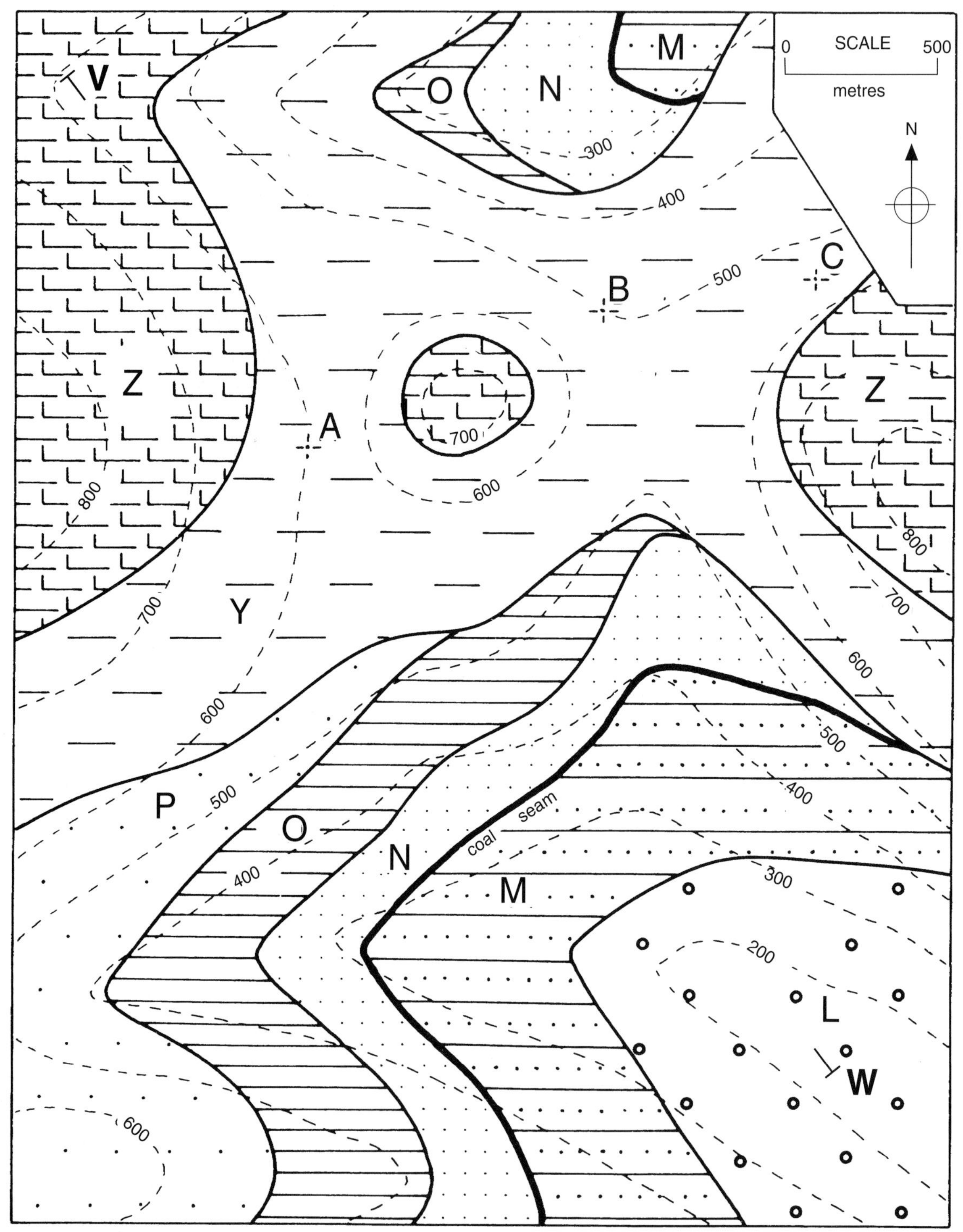

MAP 10 Find the plane of unconformity. Deduce the direction and amount of dip of the two series of beds. Draw a section along the line from V near the north-west corner of the map to W near the south-east corner. Would the coal seam be encountered in boreholes situated at points A, B and C? If the coal is present, calculate its depth below the ground surface; if it is absent, suggest an explanation for its absence. Indicate the position of the coal seam beneath bed Y.

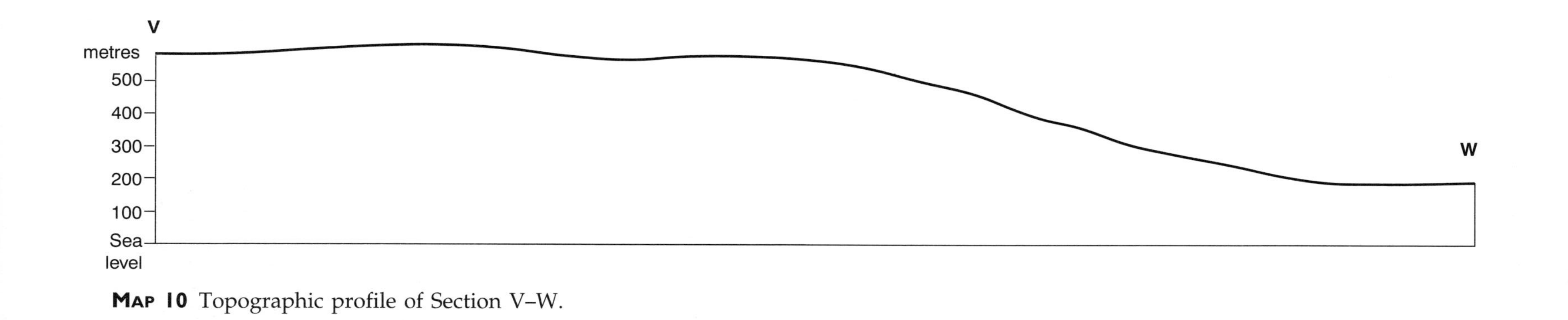

MAP 10 Topographic profile of Section V–W.

Note on Map 10. The sub-unconformity position of the coal seam may be obtained by joining with a straight line the two points where the coal seam meets the base of Bed Y (which is the plane of unconformity). To confirm that this line is correct, note where the 300 m structure contour for the coal seam intersects the 300 m structure contour on the base of bed Y. Also note where the 400 m structure contour for the coal intersects the 400 m structure contour on the base of Y. Both these points should lie on the line that you have drawn.

in practice remove all post-unconformity strata with a bulldozer and other earth-moving equipment to reveal the plane of unconformity. We can, by a study of the structure contours, deduce from a map just what we should expect to see if that was possible.

What we seek are the 'outcrops' of the older strata on the plane of unconformity. This plane can be defined by its structure contours. If we now take the structure contours drawn on the geological boundaries of the older set of strata and note where they intersect, not ground contours, but the structure contours of the plane of unconformity, we can plot – where the two sets of structure contours intersect at the same height – a number of points which will define the sub-unconformity outcrops. See Note on Map 10. The topic will be revised on a later page and Maps 12 and 17 give further practical exercises.

EXERCISES USING GEOLOGICAL SURVEY MAPS

1. **Assynt 1″ Geological Survey special sheet** Examine the western part of the map to find the unconformities at the base of the Torridonian and the base of the Cambrian. Draw a section along the north–south grid line 22 to show these unconformities. From your knowledge of the conditions under which the Torridonian and Lower Cambrian were deposited, can you explain how and why these unconformities differ?
2. **Shrewsbury 1″ Map No. 152** Examine the map carefully. How many of the unconformities indicated in the geological column in the margins are deducible from the map evidence?

4

FOLDING

We have seen that strata are frequently inclined (or dipping). On examining the strata over a wider area it is found that the inclination is not constant and, as a rule, the inclined strata are part of a much greater structure. For example, the Chalk of the South Downs dips generally southwards towards the Channel – as can be determined by examination of the 1:50,000 Geological Survey map of Brighton (Sheet No. 315). We know, however, that in the North Downs the Chalk dips to the north (passing beneath the London Basin); the inclined strata of the South and North Downs are really parts of a great structure which arched up the rocks including the Chalk over the Wealden area. Not all arching of the strata is of this large scale and minor folding of the strata can be seen to occur near the centre of the Brighton sheet.

Folding of strata represents a shortening of the earth's crust and results from compressive forces.

ANTICLINES AND SYNCLINES

Where the beds are bent upwards into an arch the structure is called an anticline (*anti* = opposite: *clino* = slope; the beds dip away from each other on opposite sides of the arch-like structure). Where the beds are bowed downwards the structure is called a syncline (*syn* = together: *clino* = slope; beds dip inwards towards each other) (Fig. 12). In the simplest case the beds on each side of a fold structure, i.e. the limbs of the fold, have the same amount of dip and the fold is symmetrical. In this case a plane bisecting the fold, called the axial plane, is vertical. The fold is called an upright fold whenever the axial plane is vertical or steeply dipping. (Where axial planes have a low dip or are nearly horizontal, see Fig. 43(a), folds are called flat folds.)

The effect of erosion on folded strata is to produce outcrops such that the succession of beds of one limb is repeated, though of course in the reverse order, in the other limb. In an

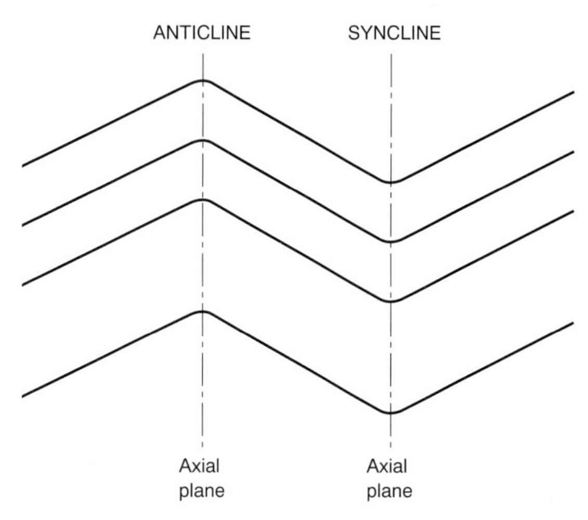

FIGURE 12 Diagrammatic section of folded strata.

eroded anticline the oldest bed outcrops in the centre of the structure and, as we move outwards, successively younger beds are found to outcrop (Fig. 13). In an eroded syncline, conversely, the youngest bed outcrops at the centre of the structure with successively older beds outcropping to either side (Fig. 14).

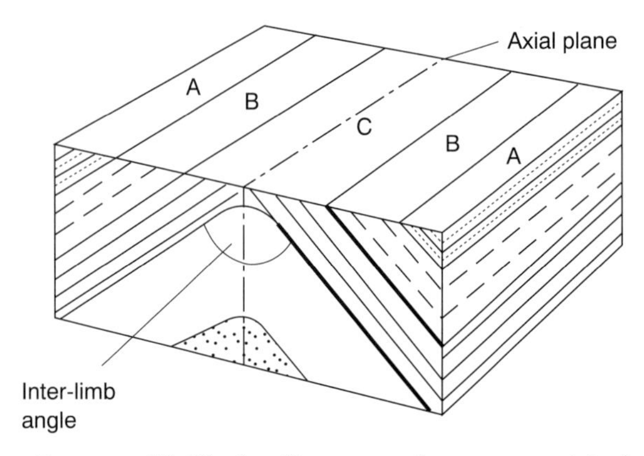

FIGURE 13 Block diagram of a symmetrical anticline (an upright fold).

ASYMMETRICAL FOLDS

In many cases the stresses in the earth's crust producing folding are such that the folds are not symmetrical like those described above. If the beds of one limb of a fold dip more steeply than the beds of the other limb, then the fold is asymmetrical. The differences in dip of the beds of the two limbs will be reflected in the widths of their outcrops, which will be narrower in the case of the limb with the steeper dip (Fig. 15). (See also p. 11 and Fig. 8). Now, the axial plane bisecting the fold is no longer vertical but is inclined and the fold is called an inclined fold.

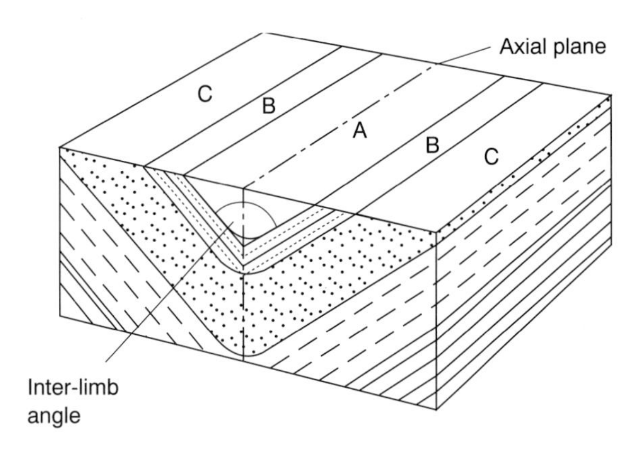

FIGURE 14 Block diagram of a symmetrical syncline (an upright fold).

Overfolds

If the asymmetry of a fold is so great that both limbs dip in the same direction (though with different angles of dip), that is to say the steeply dipping limb of an asymmetrical fold has been pushed beyond the vertical so that it has a reversed – usually steep – dip, the fold is called an overfold (Fig. 16). The strata of the limb with the reversed dip, it should be noted, are upside down, i.e. inverted.

Figures 14 to 17 show a progressive decrease in inter-limb angle.

Figures 14 to 17 illustrate fold structures produced in response to increasing tectonic stress. Further terminology should be noted. Where the limbs of a fold dip at only a few degrees it is a gentle fold; with somewhat greater dip (Figs 13, 14) a fold is described as open; with steeper dipping limbs a fold is described as closed; and with parallel limbs (Fig. 17(a),(b)) the folding is isoclinal.

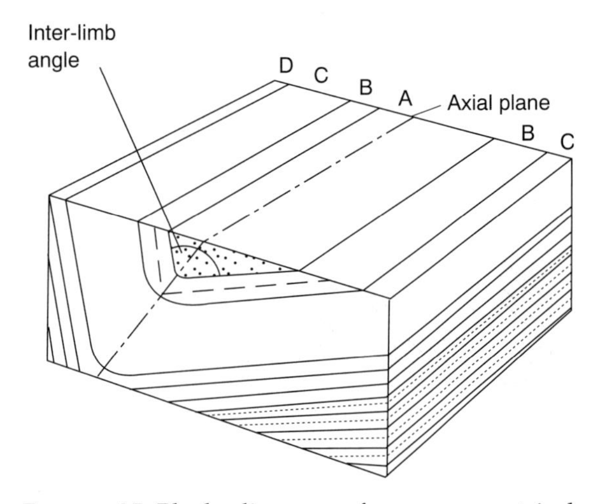

FIGURE 15 Block diagram of an asymmetrical syncline (an inclined fold).

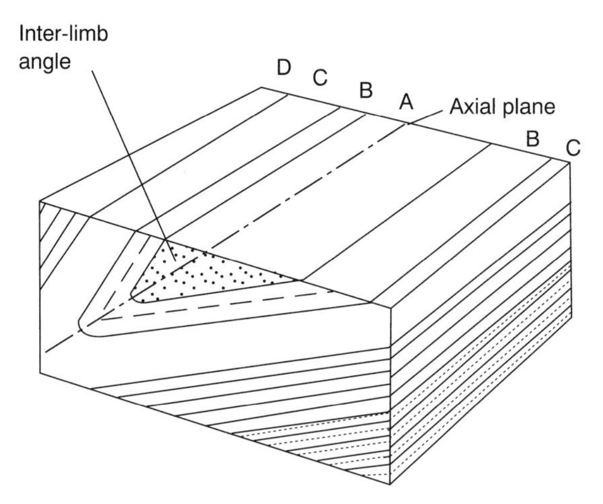

FIGURE 16 Block diagram of an overfold (an overfolded syncline).

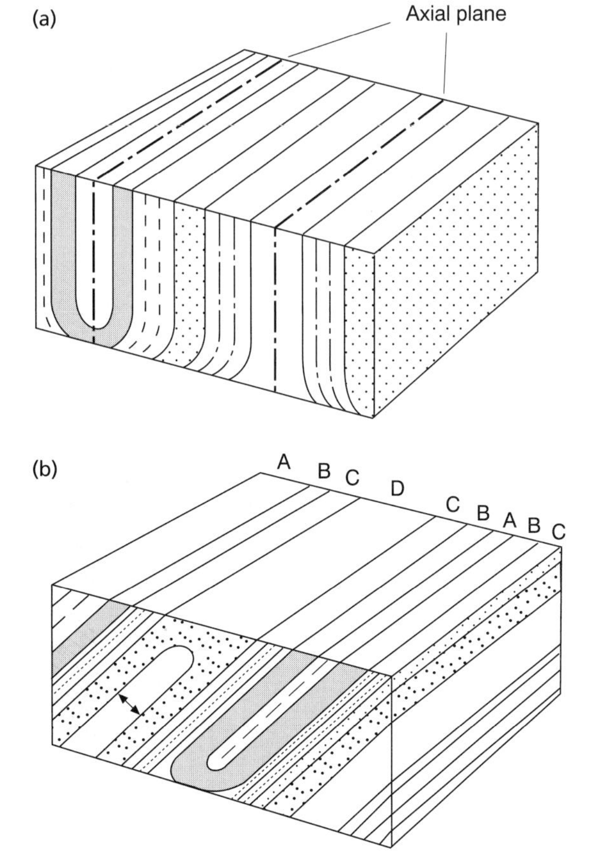

FIGURE 17 Block diagrams of isoclinal folding: (a) upright folds; (b) overfolds.

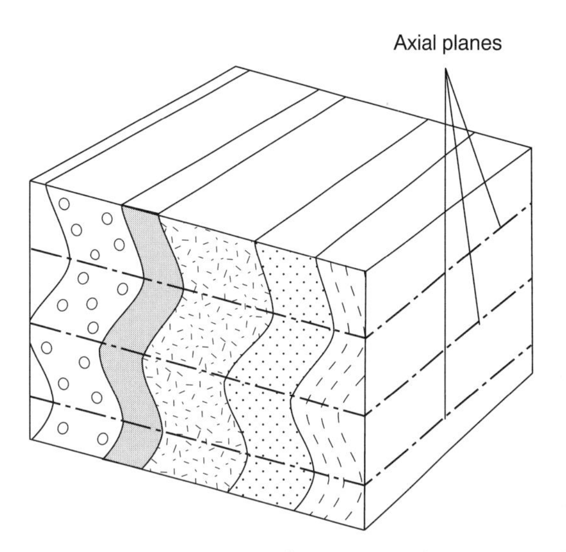

FIGURE 18 In an area of more complex structural geology, not produced simply by one epoch of compression, flat folds such as these may be found. They are open folds with horizontal or near-horizontal axial planes.

Isoclinal folds

Isoclinal folds are a special case of overfolding in which the limbs of a fold both dip in the same direction at the same angle (*isos* = equal: *clino* = slope), as the term suggests (Fig. 17(b)). The axial planes of a series of such folds will also be approximately parallel over a small area, but over a larger area extending perhaps forty kilometres (greater than that portrayed in a problem map) they may be seen to form a fan structure.

SIMILAR AND CONCENTRIC FOLDING

When strata, originally horizontal, are folded it is clear that the higher beds of an anticline form a greater arc than the lower beds (and the converse applies in a syncline). Theoretically, at least two mechanisms are possible: the beds on the outside of a fold may be relatively stretched while those on the inside are compressed, or

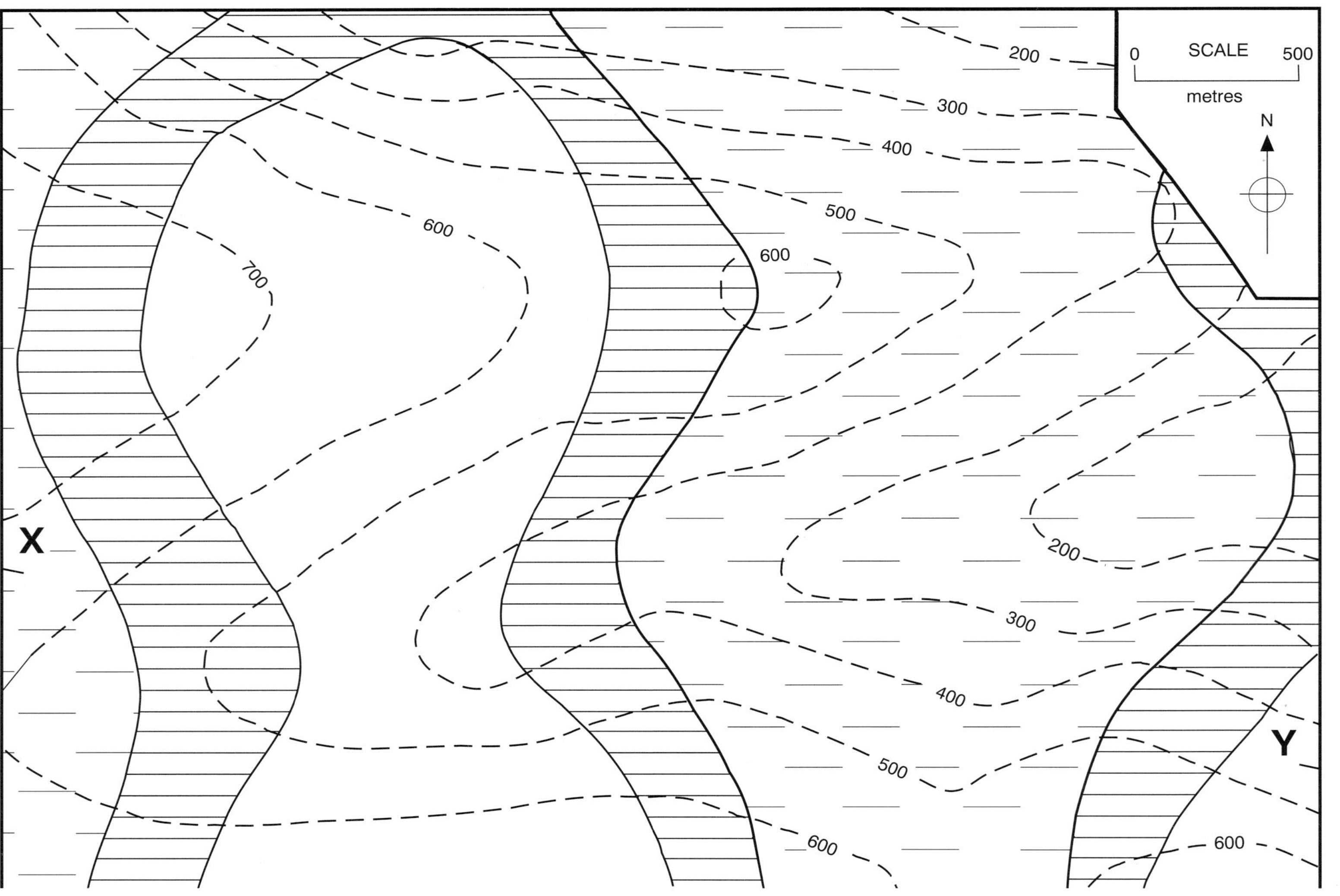

MAP 11 Draw structure contours for the upper and lower surface of the shaded bed of shale. Is the direction of strike approximately north–south or east–west? Indicate on the map the position of an anticlinal axial trace and the position of a synclinal axial trace. Draw a section along the line X–Y. (The axial trace is the outcrop of the axial plane.)

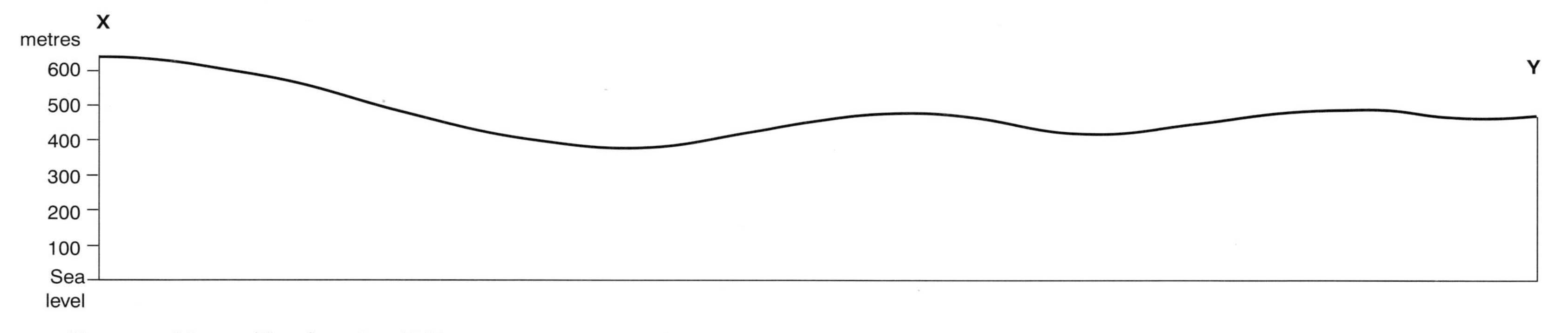

Topographic profile of section X–Y.

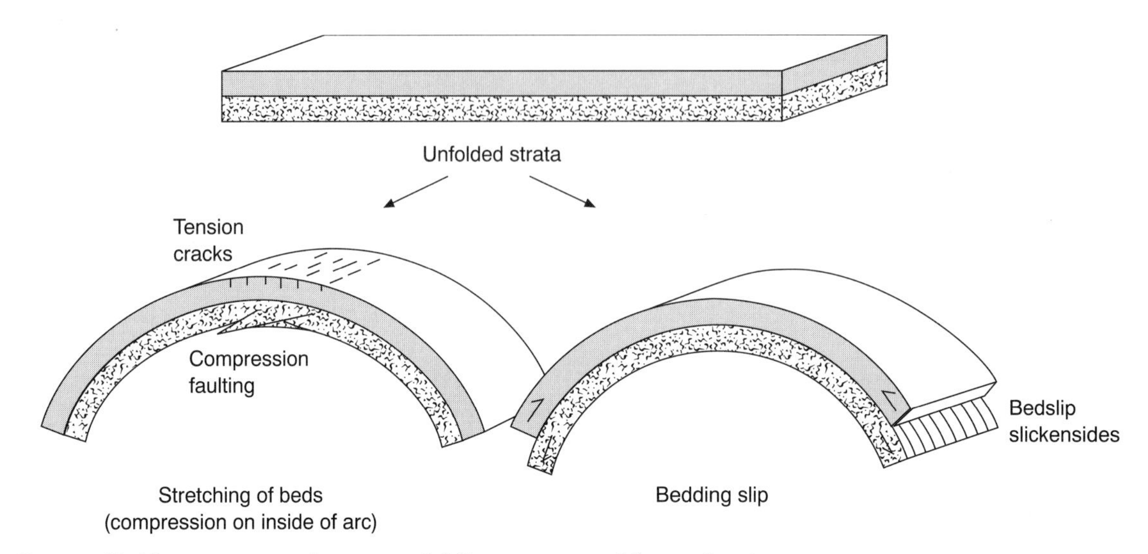

FIGURE 19 The response of strata to folding: two possible mechanisms.

the beds on the outside of a fold may slide over the surface of the inner beds (Fig. 19).

The way in which beds will react to stress depends upon their constituent materials (and the level in the crust at which the rocks lie). Competent rocks such as limestone and sandstone do not readily extend under tension or compress under compressive forces but give way by fracturing and buckling while incompetent rocks such as shale or clay can be stretched or squeezed. Thus in an alternating sequence of sandstones and shales the sandstones will fracture and buckle while the shales will squeeze into the available spaces.

Concentric folds

The beds of each fold are approximately concentric, i.e. successive beds are bent into arcs having the same centre of curvature. Beds retain their constituent thickness round the curves and there is little thinning or attenuation of beds in the limbs of the folds (Fig. 20(a)).

Straight limbed folds also maintain the uniformity of thickness of the beds (except in the hinge of the fold) and folding takes place by slip along the bedding planes as it does in the case of concentric folds. Although typically developed in thinly bedded rocks (such as the Culm Measures of North Devon) most of the problem maps in this book that illustrate folding have straight limbed folds since these provide a simple pattern of equally spaced structure contours on each limb of a fold.

Similar folds

The shape of successive bedding planes is essentially similar, hence the name (Fig. 20(b)). Thinning of the beds takes place in the limbs of the folds (and a strong axial plane cleavage is usually developed). This type of folding probably occurs when temperatures and pressures are high.

Two possible directions of strike

A structure contour is drawn by joining points at which a geological boundary surface (or bedding plane) is at the same height. By definition this surface is at the same height along the whole length of that structure contour. Clearly, if we join points X and Y (Fig. 21) we are constructing a structure contour for the bedding plane shown, for not only are points

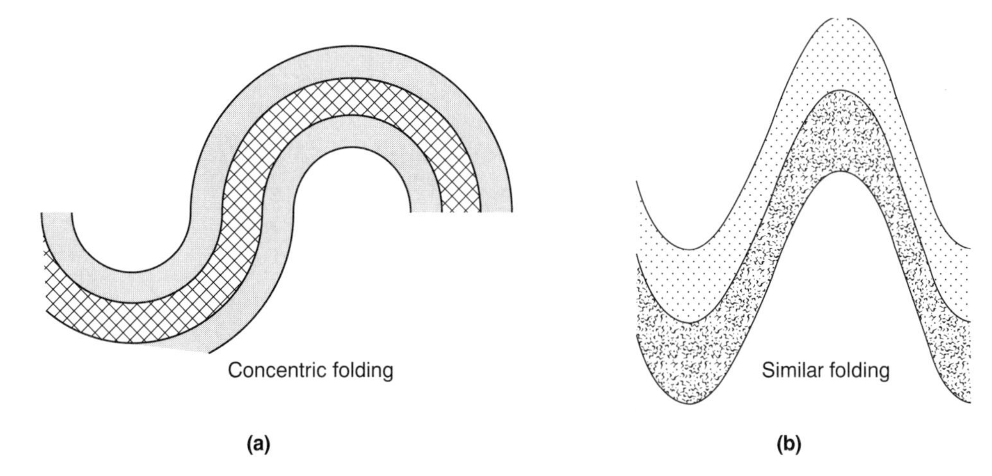

FIGURE 20 The shape of concentric and similar folding seen in section.

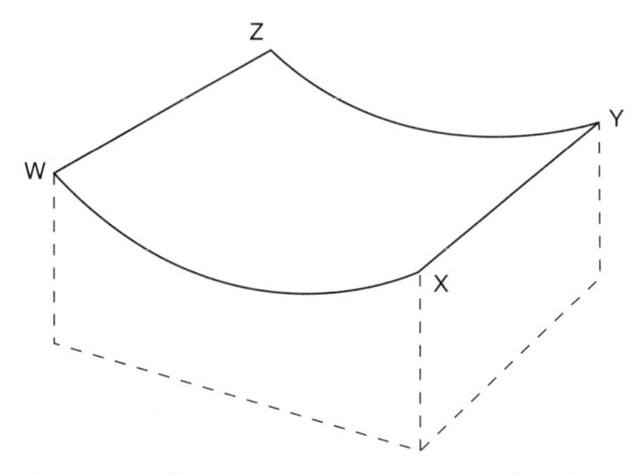

FIGURE 21 Block diagram to illustrate the direction of structure contours in folded strata.

X and Y at the same height but the bedding plane is at the same height along the line X–Y. If, however, we join the points W and X, although they are at the same height, we are not constructing a structure contour, for the bedding plane is not at the same height along the line W–X; it is folded downwards into a syncline. Thus, if we attempt to draw a structure contour pattern that proves to be incorrect, we should look for the correct direction approximately at right angles to our first attempt. It should also be noted that an attempt to visualize the structures must be made. For example, in Map 11 the valley sides provide, in essence, a section which suggests the synclinal structure, especially if the map is turned upside down and viewed from the north. Map 13 similarly reveals the essential nature of the structures by regarding the northern valley side as a section. To facilitate this, fold the map at right angles along the line of the valley bottom. Now regard the top half of the map as an approximate geological section.

What is the test of whether we have found the correct direction of strike? In these relatively simple maps the structure contours should be parallel and equally spaced (at least for each limb of a fold structure). Furthermore, calculations of true thicknesses of a bed at different points on the map should give the same value.

SECTIONS ACROSS PUBLISHED GEOLOGICAL SURVEY MAPS

Lewes, 1:50,000, BGS Map (Sheet No. 319) Solid & Drift edition 1979

Study the geology of this area on the southern limb of the Wealden anticline. Note the general structural trend, close to E–W, and the 'younging' of beds southwards. Study the relationship of topography to geology. Draw a section along the line of Section 1 (south of grid line 117). Make the vertical exaggeration ×4 (the same as the BGS section on this map).

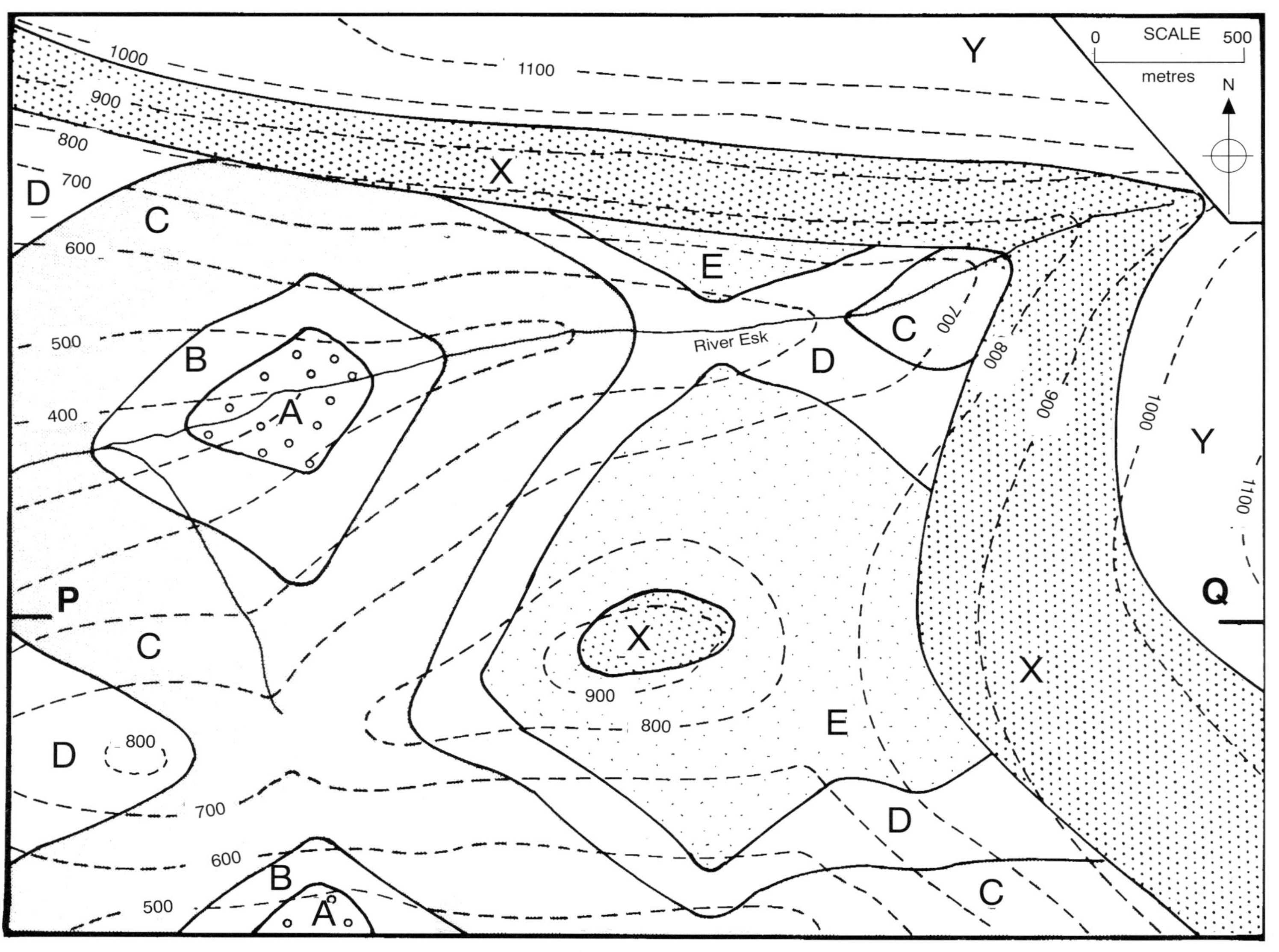

MAP 12 Indicate on the map the outcrop of the plane of unconformity. Insert the axial traces of the folds. Draw a section along the line P–Q to illustrate the geology. Indicate on the map the extent of Bed E beneath the overlying strata.

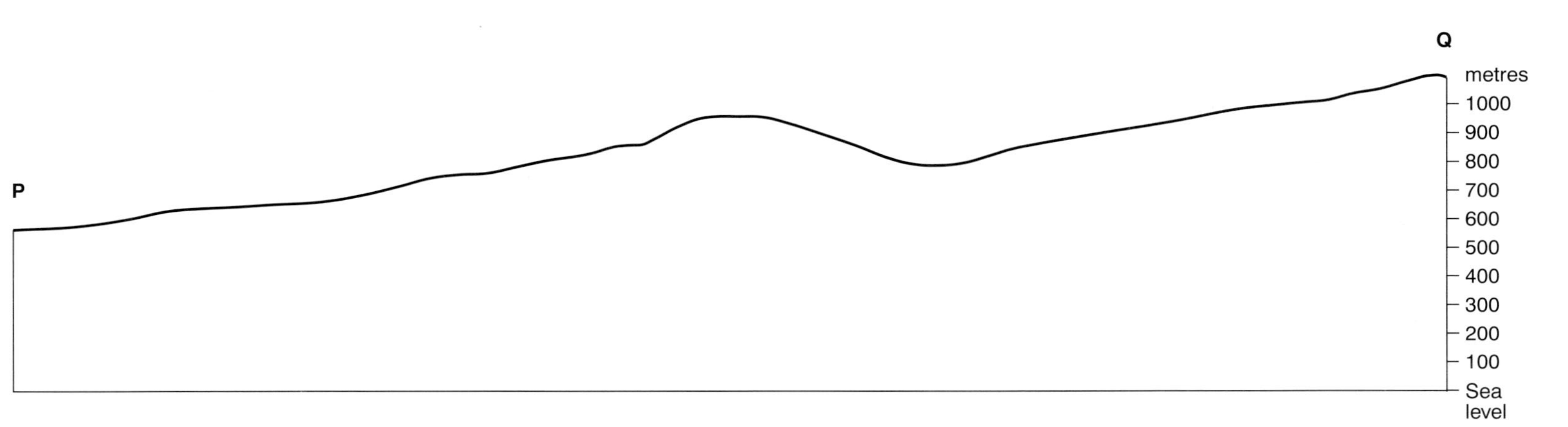

Topographic profile of section P–Q.

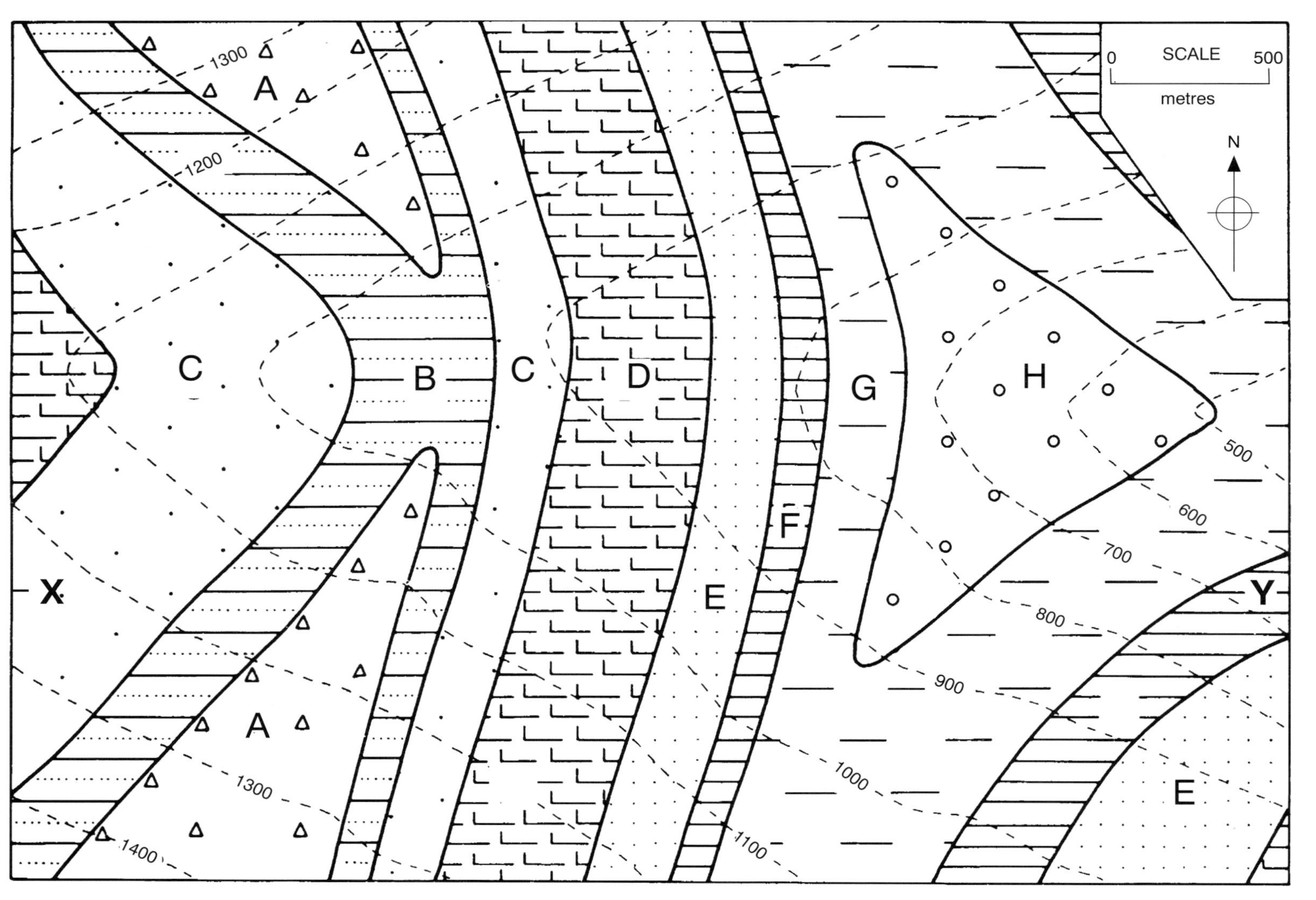

Map 13 Draw structure contours on all the geological boundaries and deduce dips and strikes. What type of folds are these? Note that all geological boundaries 'V' downstream. Draw a section along the east–west line X–Y. Draw in the axial traces, i.e. the outcrops of the axial planes of the folds.

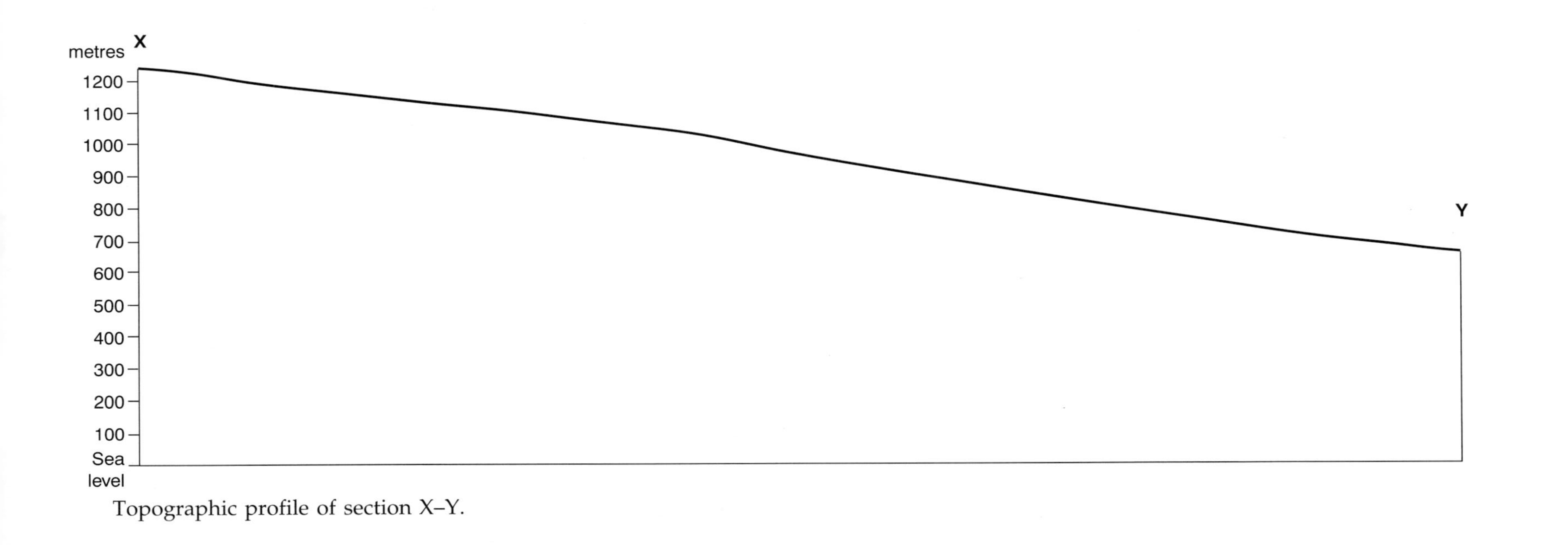

Topographic profile of section X–Y.

Note on Map 12 You should find three areas where Bed E occurs underneath younger beds. The eastern extent of Bed E could be deduced by joining the three points where the D/E junction is cut by the base of Bed X. However, to confirm that this is correct it is necessary to use the intersection of the two sets of structure contours (on D/E and on the base of X). This method is the only way in which the western extent of Bed E can be defined.

Remember (see p. 24 'Sub-unconformity outcrops') that the topography is irrelevant to the solution of this part of the problem. The surface on which we are plotting the outcrop of the D/E boundary is the plane of unconformity, defined by the structure contours drawn on the base of Bed X. You will find it necessary to use intersections of structure contours which are now above ground level: although the strata in question have now been eroded away from these points, the points themselves remain valid as construction points.

Notes on the Lewes map Here, as elsewhere in the south-east of England, the Chalk of the Upper Cretaceous age is a compact rock more resistant to erosion than the softer clays and sands of the pre-Chalk age. There is less information on the dip of strata than desirable, although dips are given to the north and south of the town of Lewes. Remember to allow for the vertical exaggeration by multiplying the gradient (tangent) of given dips by 4. Where dip information is less than adequate the beds must be fitted to the outcrops at an angle of dip that gives the correct thicknesses of strata shown in the stratigraphic column on the map.

5

FAULTS

The previous chapter showed how geological strata subjected to stress may be bent into different types of fold. Strata may also respond to stress (compression, extension or shearing) by fracturing.

Faults are fractures which displace the rocks. The strata on one side of a fault may be vertically displaced tens, or even hundreds, of metres relative to the strata on the other side. In another type of fault the rocks may have been displaced horizontally for a distance of many kilometres. While in nature a fault may consist of a plane surface along which slipping has taken place, it may on the other hand be represented by a zone of breciated rock (i.e. rock composed of angular fragments). For the purposes of mapping problems it can be treated as a plane surface, usually making an angle with the vertical.

All structural measurements are made with reference to the horizontal, including the dip of a fault plane. This is a measure of its slope (cf. the dip of bedding planes, etc.). The term 'hade', formerly used, is the angle between the fault plane and the vertical (and is therefore the complement of the dip). It causes some confusion but will be widely encountered in books and on maps and is, for this reason, shown in Figs 22 and 23. It will, no doubt, gradually drop out of usage.

The most common displacement of the strata on either side of a fault is in a vertical sense. The vertical displacement of any bedding plane is

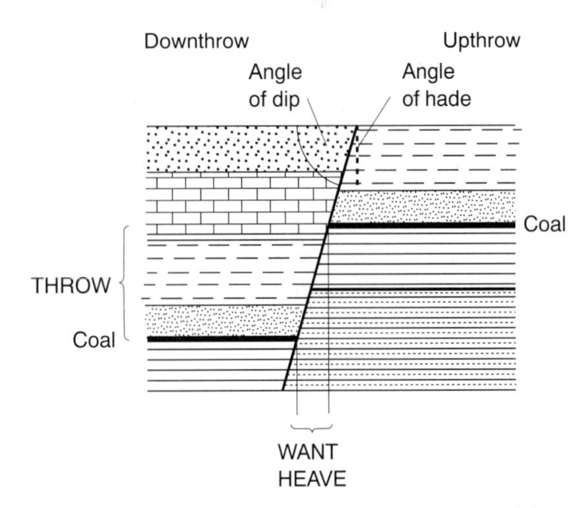

FIGURE 22 Section through strata displaced by a normal fault (after erosion has produced a near-level ground surface).

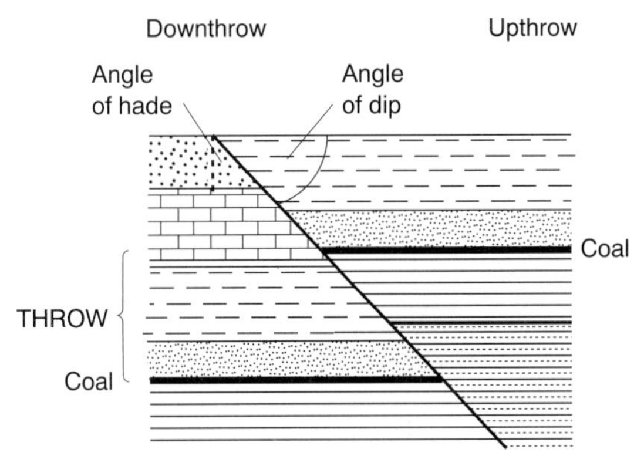

FIGURE 23 Section through strata displaced by a reversed fault.

called the throw of the fault. Other directions of displacement are dealt with later in this chapter.

Normal and reversed faults

If the fault plane is vertical or dips towards the downthrow side of a fault, it is called a normal fault (Fig. 22). If the fault plane dips in the opposite direction to the downthrow (i.e. towards the upthrow side) it is a reversed fault (Fig. 23).

In nature, the dip of a reversed fault is generally lower than that of a normal fault. It may be less than 45°. In an area of strong relief the outcrop of a reversed fault may be sinuous. The outcrop of a normal fault (commonly with a dip in the 85-75° range, but as low as 50° in some examples) will usually be much straighter – unless the fault plane itself is curved (see Map 15). Of course, where the area is of low relief, the outcrop of a fault plane, normal or reversed, will be virtually straight.

On some geological maps, such as those produced in Canada and the United States, the direction of dip of fault planes is shown. The angle of dip of the fault plane may be given in the map description. British Geological Survey maps show the direction of the throw of faults by means of a tick on the downthrow side of the fault outcrop.

Any sloping plane, including a fault plane, can be defined by its structure contours. It is possible on both Maps 14 and 16 to construct contours for the fault plane. The method is exactly the same as for constructing structure contours on bedding planes, described in Chapter 1. From these structure contours the direction of dip of the fault plane can be deduced and then, by reference to the direction of downthrow, it can be deduced whether the fault is normal or reversed.

The effects of faulting on outcrops

Consider the effects of faulting on the strata: those on one side of a fault are uplifted, relatively, many metres. Since this uplift is not as a rule a rapid process and the strata will be eroded away continuously, a fault may not make a topographic feature, although temporarily a fault scarp may be present (Fig. 24), especially after sudden uplift resulting from an earthquake.

Some faults which bring resistant rocks on the one side into juxtaposition with easily eroded rocks on the other side may be recognized by the presence of a fault line scarp (cf. the fault scarp resulting from the actual movement).

The strata that have been elevated on the upthrow side of a fault naturally tend to be eroded more rapidly than those on the downthrow side. This results in higher (younger) beds in the stratigraphical sequence being removed from the upthrow side of the

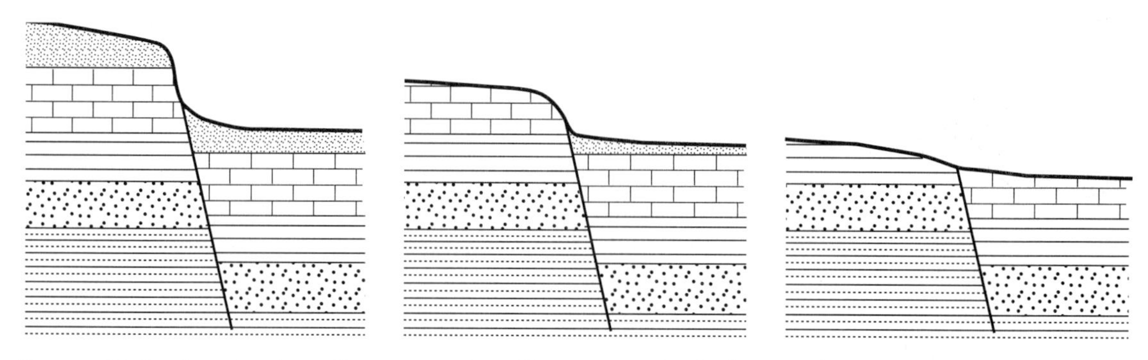

Figure 24 Sections to show the progressive elimination of a fault scarp by erosion.

fault while they are preserved on the downthrow side. It follows that we can usually determine the direction of downthrow of a fault, whether normal or reversed. Following the line of a fault across a map, there will be points where a younger bed on one side of the fault is juxtaposed against an older bed on the other side of the fault. The younger bed will be on the downthrow side of the fault.

A fault dislocates and displaces the strata. The effect of this, in combination with erosion, is to cause discontinuity or displacement in the outcrops of the strata.

CLASSIFICATION OF FAULTS

Faults may be categorized in two ways.

1. Faults may be classified according to the direction of displacement of the blocks of strata on either side of the fault plane. So far, we have considered normal and reversed faults with a vertical displacement called throw. Movement in these faults was in the direction of dip of the fault plane. They are called dip–slip faults because the movement – or displacement – was parallel to the direction of dip of the fault plane.

   On a later page faults with lateral displacement, wrench faults, are described. Here, displacement is parallel to the strike of the fault plane and they can be described as strike–slip faults.

   In nature, in some faults the displacement is neither dip–slip nor strike–slip but oblique. Naturally, the displacement will have a vertical component (throw) and a horizontal component (lateral displacement). Such a fault may be called an oblique-slip fault (Fig. 25).

2. A different classification of faults is dependent upon their geographical pattern, especially in relation to the dip directions of the strata cut by the faults. Where the faulting is parallel or nearly so to the direction of dip of strata, the faults are called dip faults. Where the faulting is more or less at right angles to the direction of dip of strata, i.e. approximately parallel to their strike, the faults are called strike faults. Examples of both are found on Problem Map 15. Faults that are neither in the dip direction nor the strike direction may be called oblique faults.

It is particularly important not to confuse the two schemes of classification discussed above: the terms are unfortunately very similar. Test yourself: what is a dip–slip strike fault? Figures

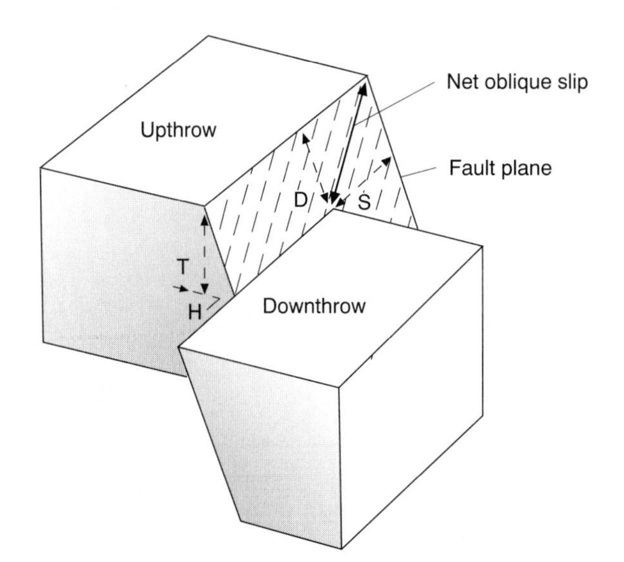

T = Throw
H = Heave
S = Strike–slip component
D = Dip–slip component

FIGURE 25 An oblique-slip fault.

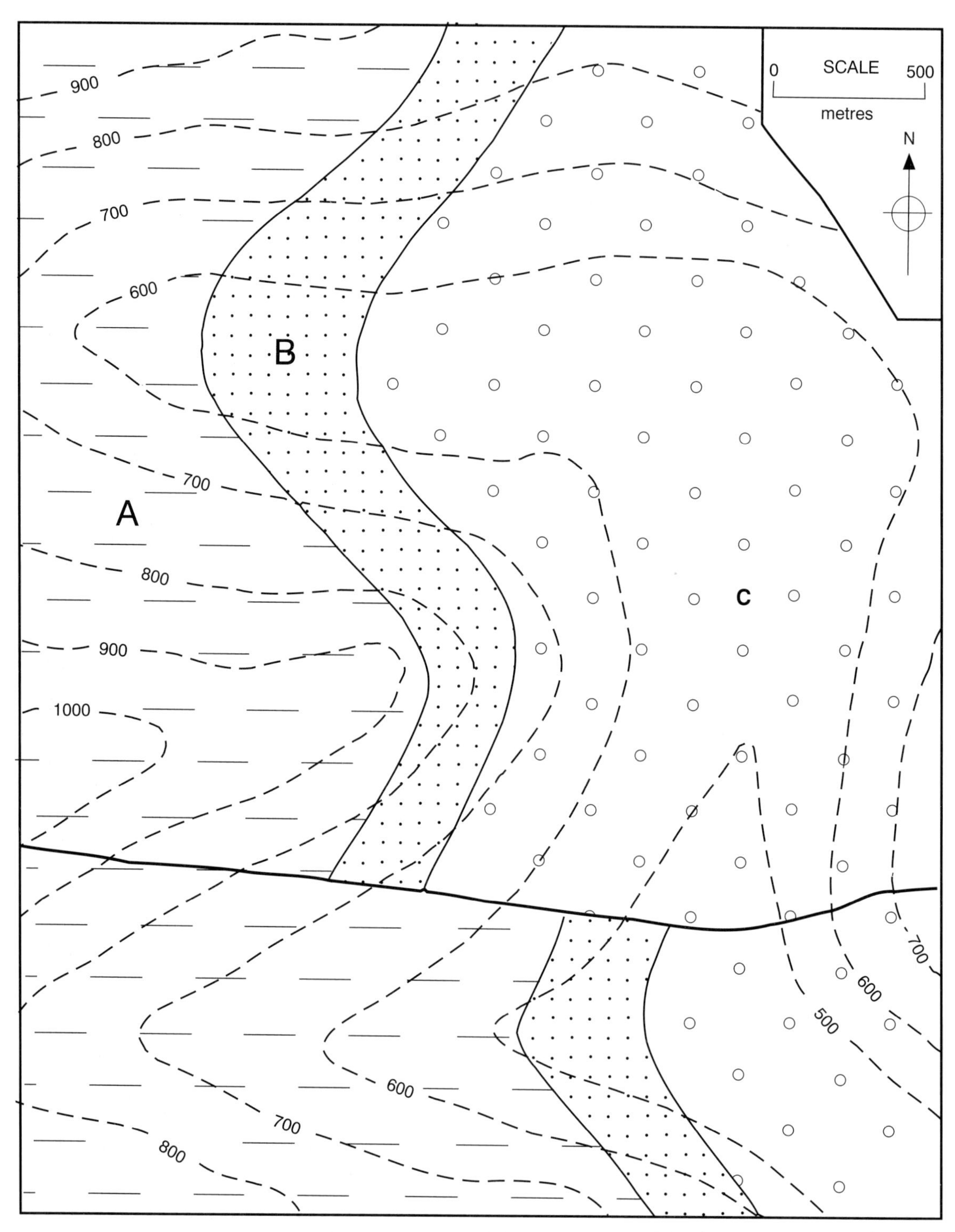

MAP 14 Draw structure contours for the upper and lower surfaces of the sandstone (stippled). What is the amount of the throw of the fault? Draw structure contours on the fault plane. Is it a normal or a reversed fault? What is the thickness of the sandstone?

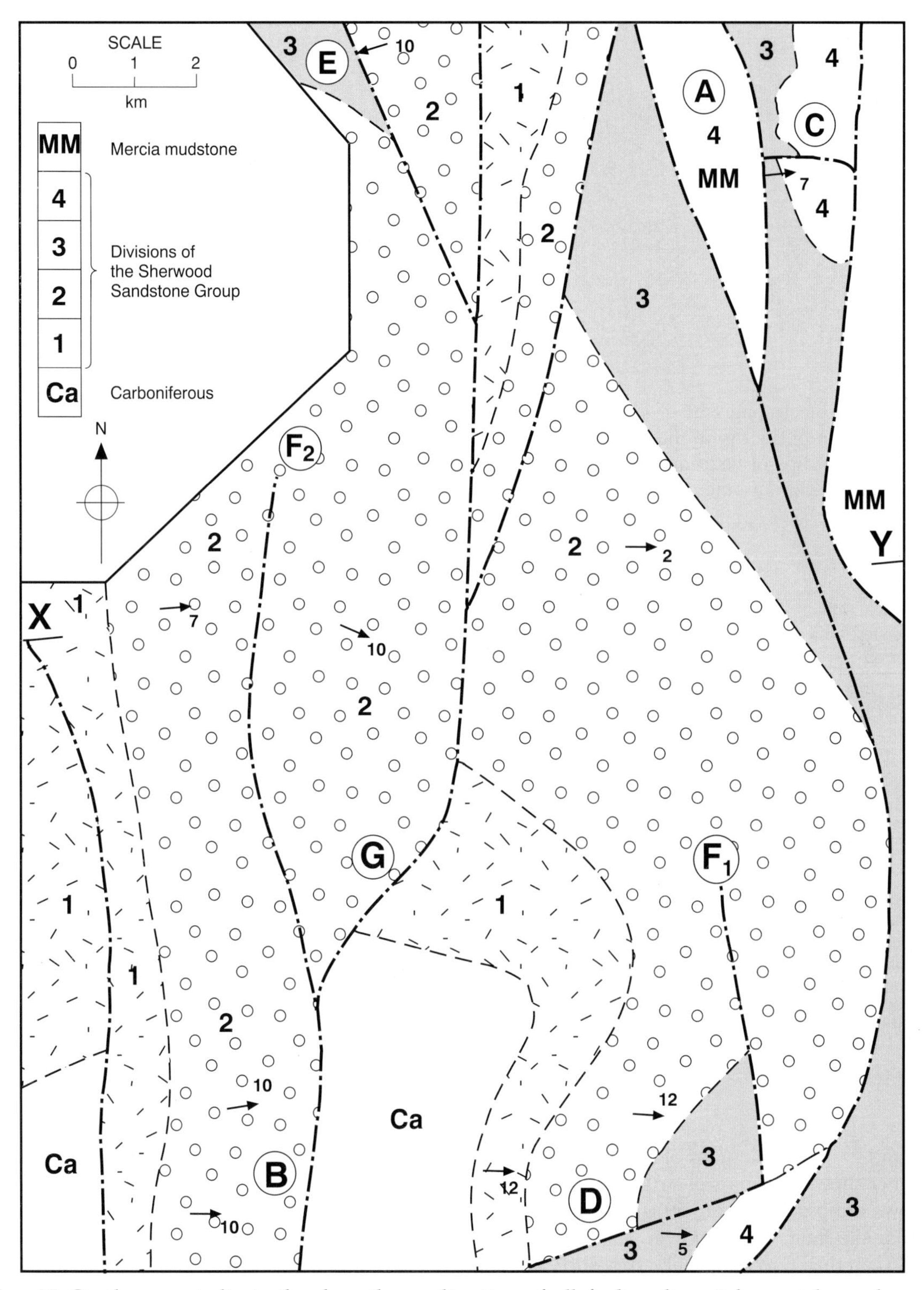

MAP 15 On the map indicate the downthrow direction of all faults where it has not been shown. Indicate examples of (a) a graben and (b) step faulting. Draw a section along the nearly east–west line X–Y to illustrate the geology. Reproduced by permission of the Director General, British Geological Survey: NERC copyright reserved.

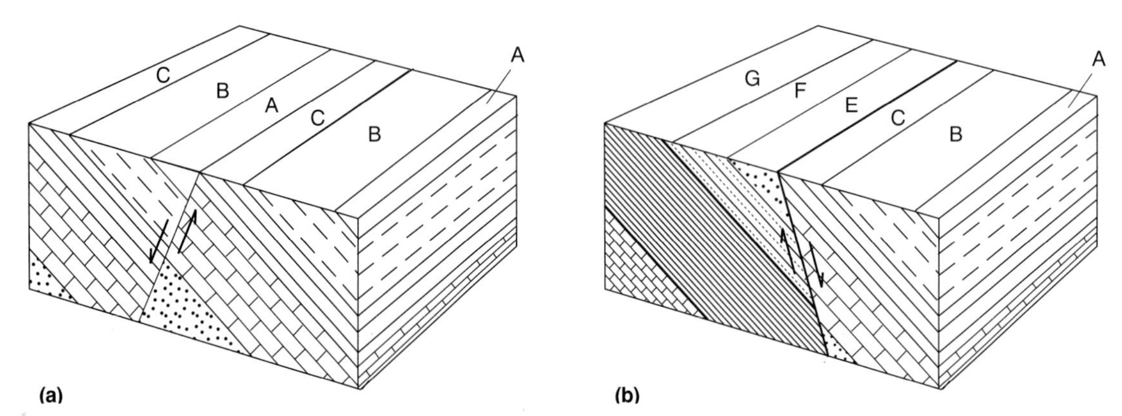

FIGURE 26 Block diagrams of a normal strike fault (a) with the direction of bedding dip opposite to the direction of the dip of the fault plane, causing repetition of part of the succession of outcrops and (b) with the dips of bedding and the fault plane in the same direction, causing a suppression of part of the succession of outcrops.

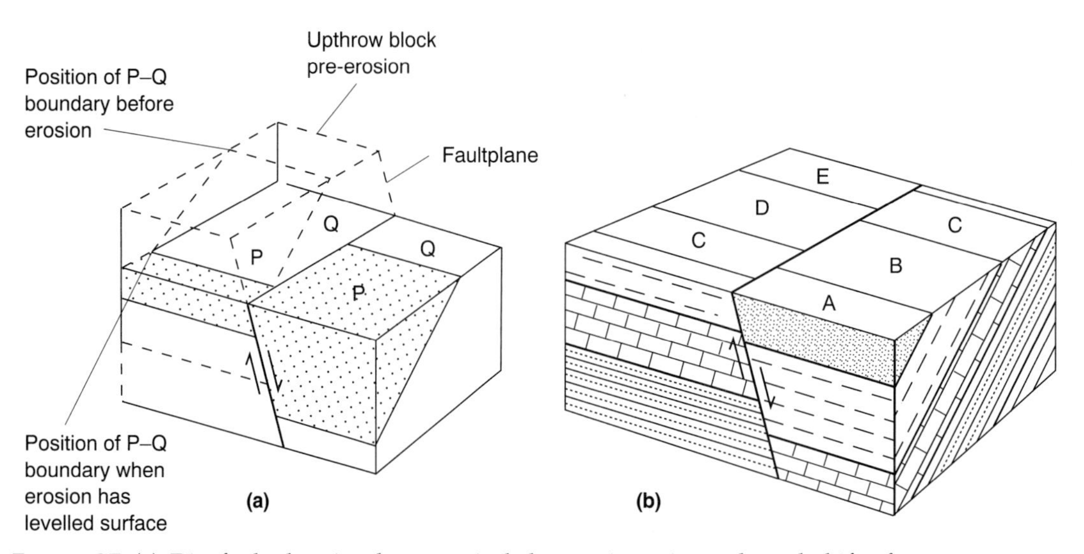

FIGURE 27 (a) Dip fault showing how vertical throw gives rise to lateral shift of outcrop as a result of erosion. The outcrops shift progressively down dip as erosion lowers the ground surface. (b) Block diagram of a normal dip fault. Note the lateral shift of outcrops although the actual displacement is vertical.

26 and 27 show examples of dip–slip faults; the former shows two cases of strike faults, the latter shows a dip fault. Figure 29 is an example of a strike–slip fault. It is, however, also a dip fault.

To return to normal dip–slip faults, sequences of outcrops encountered on a traverse may be partly repeated or may be partly suppressed.

Where the fault plane is parallel to the strike of the beds we see either repetition of outcrops (Fig. 26(a)) where the succession of beds at the surface reads A, B, C, A, B, C or the suppression of outcrops (Fig. 26(b)) where the succession of beds at the surface reads A, B, C, E, F, G.

Where the fault plane is parallel to the dip direction of the strata (a dip fault), i.e. at right angles to the strike, a lateral shift of the outcrop occurs. This must not be confused with lateral movement of the strata (see p. 46): the transposition of the outcrops is due to vertical displacement of the beds followed or accompanied by erosion which, because the strata are inclined, causes the outcrops on the upthrow side to be shifted in the direction of dip (Fig. 27).

Calculation of the Throw of a Fault

Note on Map 14 Construct the structure contours on the upper surface of the sandstone bed in the northern part of Map 14. They run north–south and are spaced at 12.5 mm intervals. Follow the same procedure for the upper surface of the sandstone south of the fault plane. The 500 m structure contour drawn on the south side of the fault, if produced beyond the fault, is seen to be coincident with the position of the 1000 m structure contour on the north side of the fault. The stratum on the south side is, therefore, 500 m lower relatively. The fault has a downthrow to the south of 500 m.

Determine the throw of the fault using the structure contours on the base of the sandstone – on each side of the fault – and check that you obtain the same value.

Shade the areas on the map where a borehole would penetrate the full thickness of the sandstone. Also shade the areas where a borehole would penetrate only a partial thickness of the sandstone. You will find a zone where the borehole would not encounter the sandstone at all due to the heave (or want) of the fault. Note that this zone will be defined by constructing structure contours for the sloping fault plane as well as for the top and bottom of the sandstone. Intersections of these two sets of lines, where they are of the same height, will define where the sandstone is cut off by the fault plane. Joining these points of intersection will define the area(s) of absence, or partial absence, of the sandstone.

Faults and Economic Calculations

It can be seen that since the fault plane dips, the intersections with a bedding plane on each side of the fault do not coincide (in plan view). Consider an economically important bed, for example of coal or ironstone. In the case of a normal fault there is a zone where a borehole

Notes on Map 15 The dip of the strata at E is anomalous, the result of a phenomenon called fault-drag.

The effect of faults which are parallel to the dip of the strata is to laterally shift outcrops (see p. 42 and Map 14). The extent of this shift depends on two factors: the amount of throw of the fault and the angle of dip of the strata. At C, on Map 15, the geological boundary is displaced very little, but at D the displacement is considerable (the shift is almost equal to the width of outcrop of Bed 4). At C and D the dips of the strata are similar so we can conclude that the fault at D has a much larger throw than the fault at C.

It may be assumed that all the faulting here is normal. Since we have no means of calculating the dip of the fault planes, in your section give faults a conventional dip towards the downthrow of 80 to 85°.

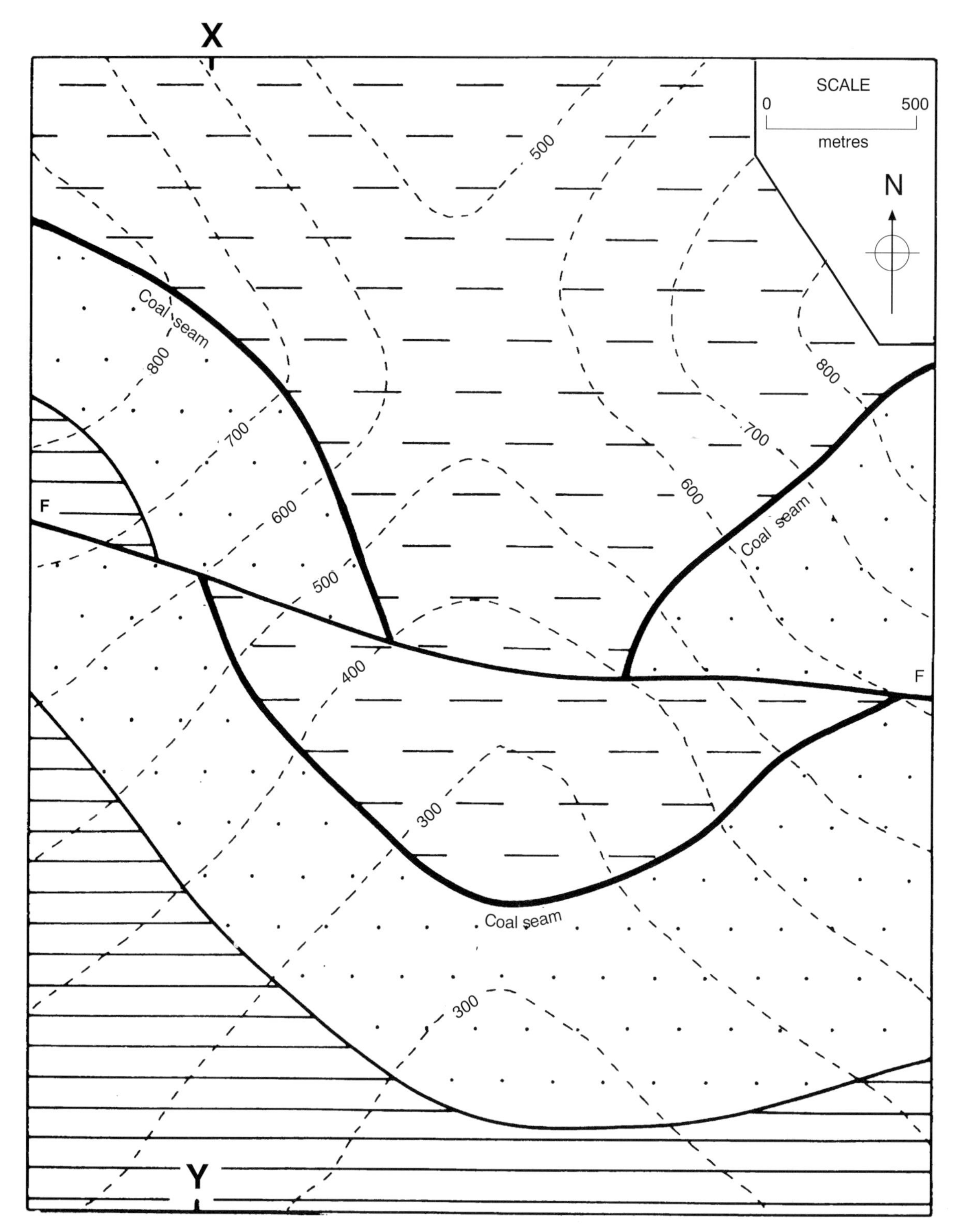

Map 16 The line F–F is the outcrop of a fault plane. The other thick line on the map is the outcrop of a coal-seam. Shade areas where coal could be penetrated by a borehole (where it has not been removed by erosion). Indicate areas in which any borehole would penetrate the seam twice. What type of fault is this? Draw a section along the line X–Y. Draw the 100 m overburden isopachyte, i.e. a line joining all points where the coal is overlaid by 100 m of strata.

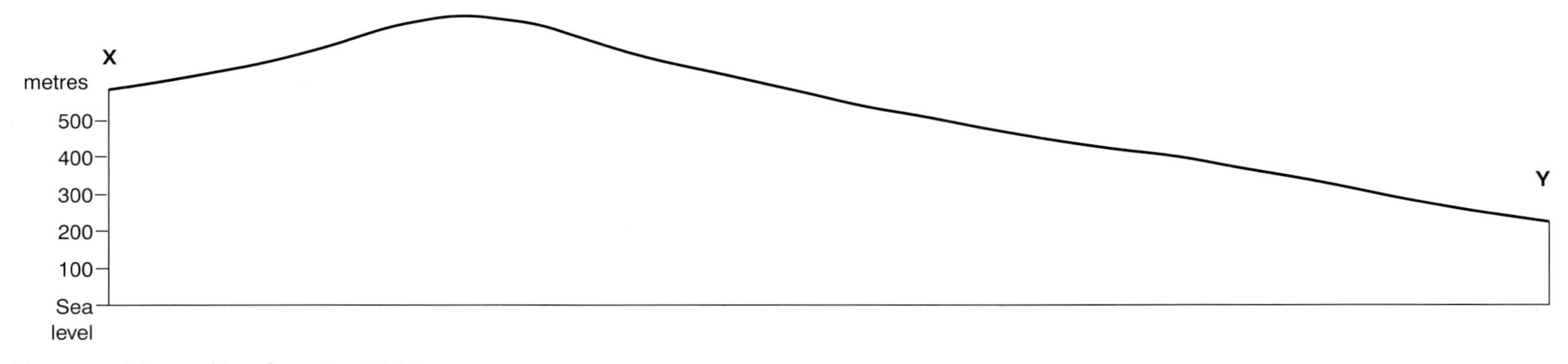

Topographic profile of section X–Y.

Notes on Map 16 The zone in which a borehole penetrates the seam twice is defined by the lines of intersection of the fault plane and the coal-seam (Fig. 28). The surfaces intersect where they are at the same height (where the coal-seam structure contours and the fault plane structure contours of the same height coincide).

would not penetrate this bed at all due to the effect of heave (see Fig. 22). This is important when calculating economic reserves, for example in a highly faulted coalfield. The estimate of reserves could be as much as 15% too great unless allowance had been made for fault heave. In the case of a reversed fault a zone exists where a borehole would penetrate the same bed twice. Figure 28 shows, in section, how a borehole after penetrating a coal seam would penetrate the fault plane and, beneath it, the same seam. In calculating reserves it is vital to recognize that it is the same seam that the borehole has encountered or reserve calculations would be wrong by a factor of approximately two.

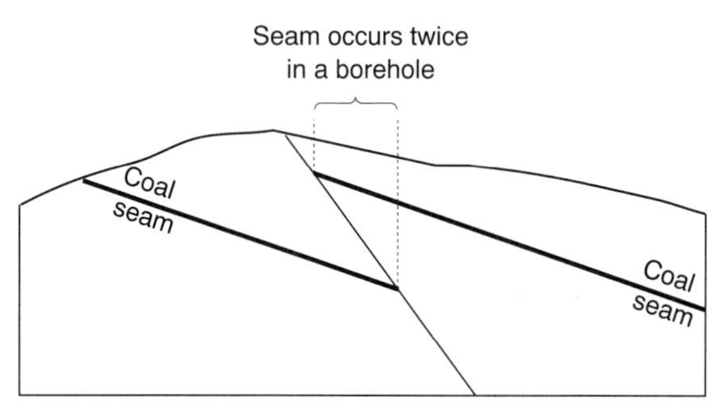

FIGURE 28 Section to show a 'low angle' (high hade) reversed fault and its importance in mining problems.

Map 15 is a slightly simplified version of the western part of the BGS map of Chester, Sheet 109 in the 1:50,000 series, solid edition. It is about half scale here. The geology of the area is relatively simple with strata dipping generally towards the east with dips in the range of 5 to 17 degrees. The topography is almost flat (and drift cover is extensive so that much of the map has been compiled using borehole data). Strata are displaced by a considerable number of faults and they display many important characteristics of faulting.

Note that faults may die out laterally; examples can be seen at F_1 and F_2. Of course, all faults die out eventually, unless cut off by another fault, as at C, though major faults may extend for tens or even hundreds of kilometres. Note also that a fault may curve, for example at G. This is not mere curvature of the outcrop of the fault plane due to the effects of topography on a sloping fault plane (see Map 16).

Most of the faults on Map 15 run approximately north–south, roughly parallel to the general strike of the strata: they can be called strike faults. The displacement is in the direction of dip of the fault planes so they can also be called dip–slip faults. Two faults are approximately parallel to the direction of dip of the strata, seen at C and D. These are, therefore, dip faults. They are also probably dip–slip faults. The Chester sheet indicates the direction of downthrow of each fault, as do most published maps, but this has been omitted from most of the faults on Map 15.

WRENCH OR TEAR FAULTS

In the case of these faults the strata on either side of the fault plane have been moved laterally relative to each other, i.e. movement has been a horizontal displacement parallel to the fault plane. In the case of simply dipping strata the outcrops are shifted laterally (Fig. 29) so that the effect, *on the outcrops*, is similar to that

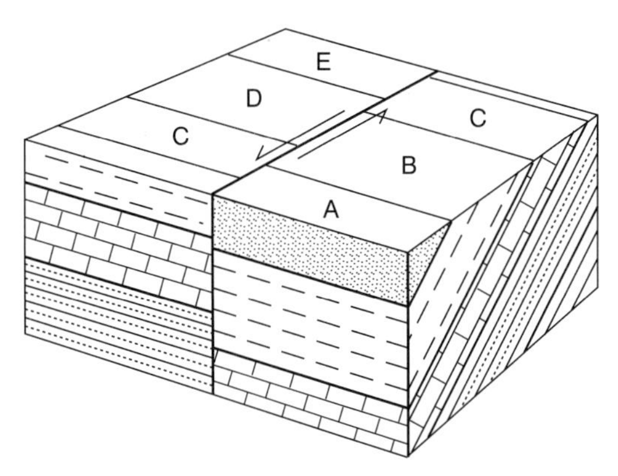

FIGURE 29 Block diagram of a wrench (= tear) fault. Note that the effect on the outcrops is similar to that of a normal dip fault (cf. Fig. 27(b)).

of a normal dip fault (cf. Fig. 27) – and in this case it is usually impossible to demonstrate strike–slip from the map alone (apparent slip only can be found).

In some geological contexts the terminology based on the direction of movement relative to the dip and strike of the fault plane is preferable. Faults in which movement has been in the strike direction of the fault plane (= strike–slip faults) include wrench faults.

PRE- AND POST-UNCONFORMITY FAULTING

After the deposition of the older set of strata, earth movements causing uplift may also give rise to faulting of the strata. The unconformable series (the younger set of beds), not being laid down until a later period, are unaffected by this faulting. Earth movements subsequent to the deposition of the unconformable beds would, if they caused faulting, produce faults that affect both sets of strata. Clearly, it is possible to determine the relative age of a fault from inspection of the geological map which will show whether the fault displaces only the older (pre-unconformity) strata or whether it displaces both sets of strata. A fault is later in age than the youngest beds it cuts.

A fault may also be dated relative to igneous intrusions, a topic dealt with in the last chapter.

STRUCTURAL INLIERS AND OUTLIERS

The increased complexity of outcrop patterns due to unconformity and faulting greatly increases the potential for the formation of outliers and inliers (these terms have been defined on p. 12). Indicate on Map 17 inliers and outliers that owe their existence to such structural features and subsequent erosional isolation.

POSTHUMOUS FAULTING

Further movement may take place along an existing fault plane. So the displacement of the strata is attributable to two or more geological periods. It follows that an older series of strata may be displaced by an early movement of the fault which did not affect newer rocks since they were laid down subsequently. The renewed movement along the fault will displace both strata so the older strata will be displaced by a greater amount since they have been displaced twice (the throws are added together).

ISOPACHYTES

Isopachytes (*iso* = equal; *pakhus* = thick) are lines of equal thickness. The simplest use of isopachytes is to show the thickness of cover material overlying a bed of economic importance, such as a coal seam or ironstone. The

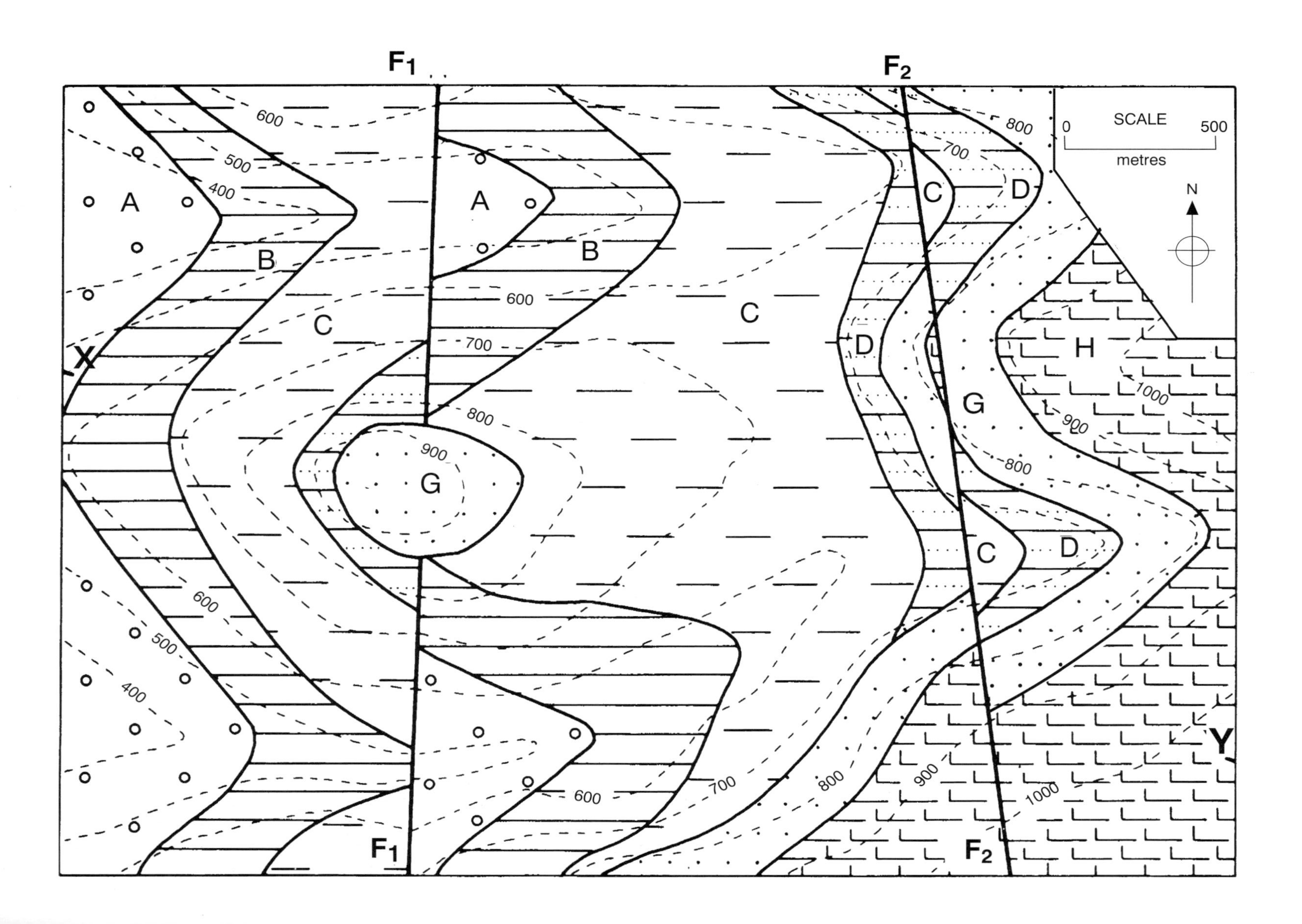

F1
F2
SCALE
0
500
metres
N
A
B
C
D
G
H
X
Y
400
500
600
700
800
900
1000

MAP 17 Calculate the direction and amount of dip of the strata below and above the plane of unconformity. Draw a section along the line X–Y. The lines F_1–F_1 and F_2–F_2 are the outcrops of two fault planes. Which fault occurred earlier in geological time? Assume that Bed B is a commercially important bed of ironstone. In order to estimate its economic potential we need to know the thickness of overlying strata which must be removed in order to mine the ironstone by opencast methods (strip mine). Draw on the map the 100 m and 200 m isopachytes for this cover (overburden).

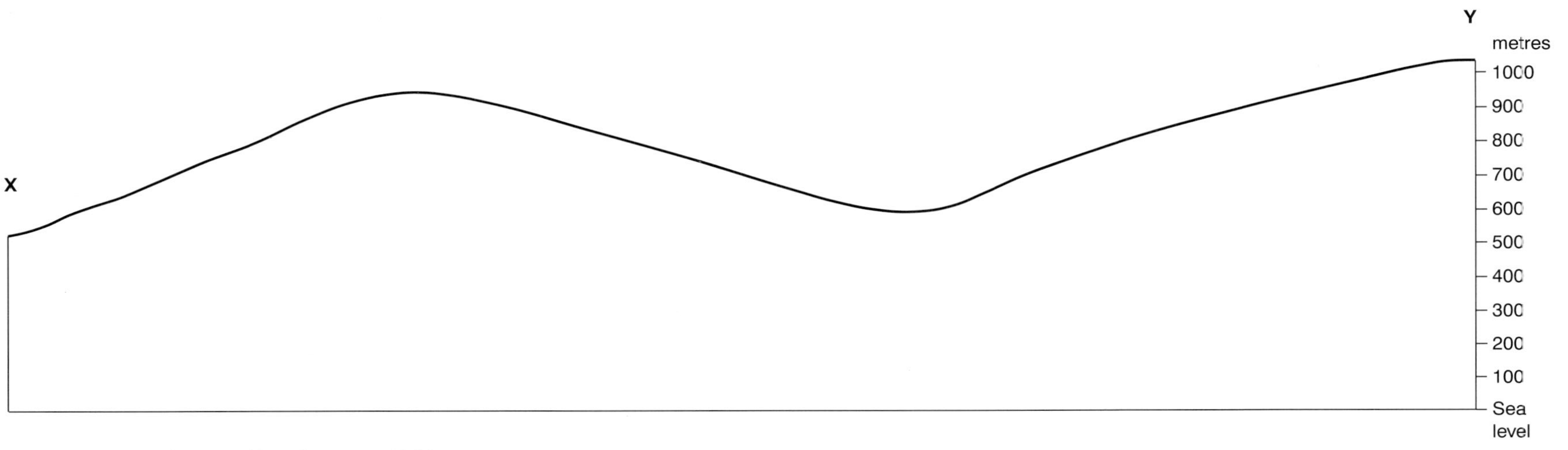

Topographic profile of section X–Y.

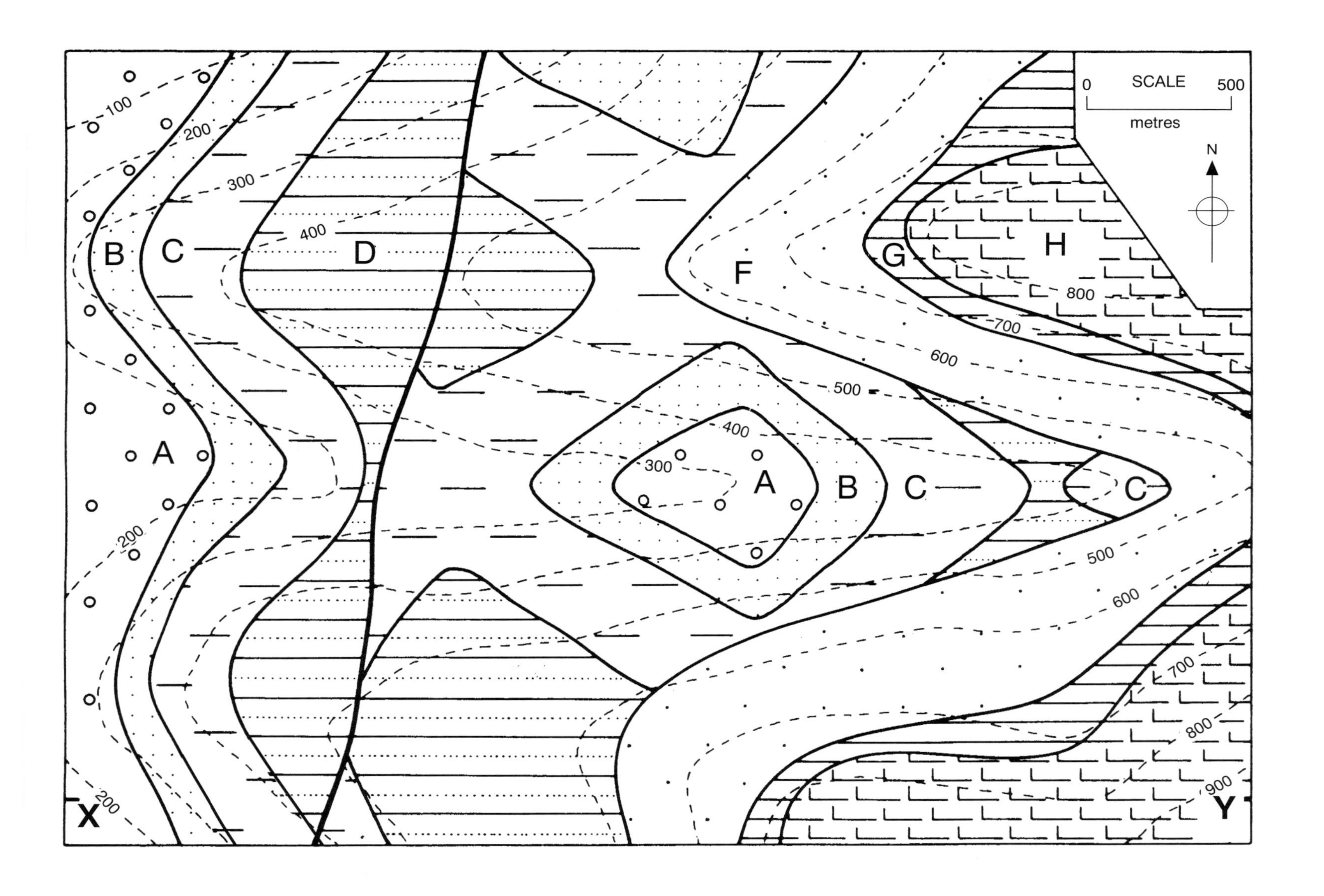

SCALE
0
500
metres
N
A
B
C
D
F
G
H
X
Y
100
200
300
400
500
600
700
800
900

Map 18 This map includes all the structural features so far introduced: folds, a fault and an unconformity. Deduce the main structural features of the area from interpretation of the outcrop patterns before attempting to construct structure contours. Write a brief geological history of the area portrayed by the map, giving the order of events producing these structural features. Draw a section along the line X–Y.

Indicate on the map where you could site a 500-m long dam to form a reservoir. Approximately how high would the dam be at its centre? What would be the approximate area of the reservoir?

Assuming that sandstone B and conglomerate A are aquifers (but the better quality of water is obtained from the sandstone) indicate where you could site a borehole for a water supply. Your simple drilling rig is limited to a depth of about 300 m.

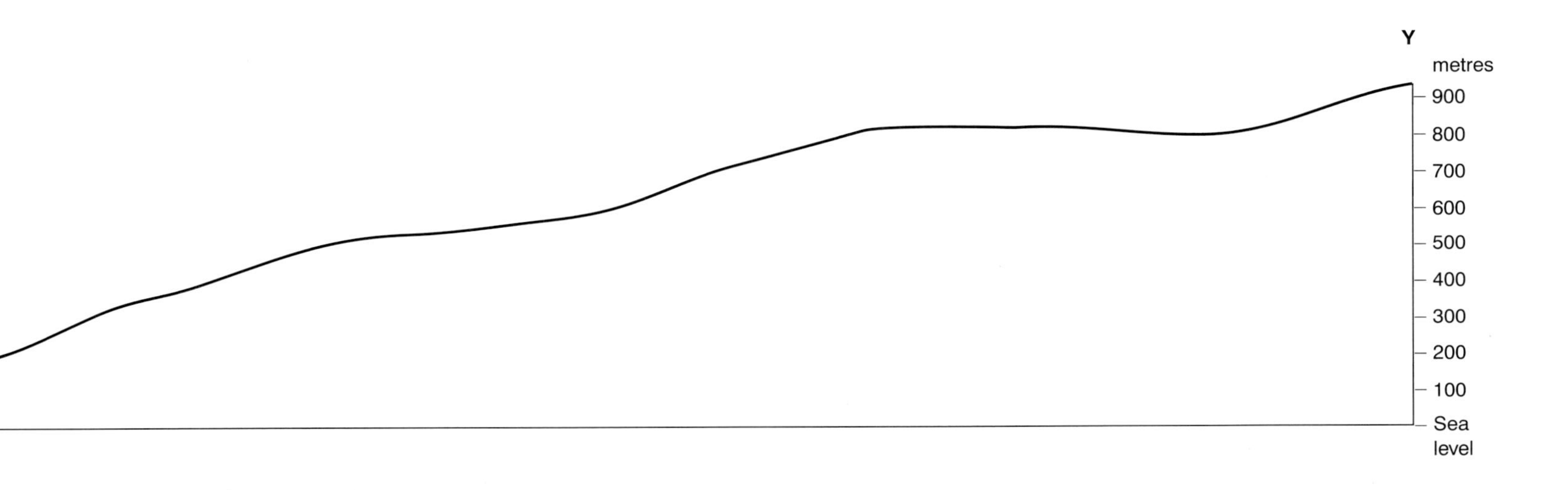

Topographic profile of section X–Y.

overlying material – whatever its composition: strata, soil or subsoil – is called the overburden. Its thickness can be determined where the height of the top of an economic bed (ironstone, Map 34) is known from its structure contours, and the height of the ground at the same point is known from the topographic contours. Wherever structure contours and topographic contours intersect on the map we can obtain a figure for the thickness of overburden (by subtracting the height of the top of the ironstone from the height of the ground). Joining up the points of equal thickness gives an isopachyte. Where ironstone and ground are at the same height the thickness of overburden is nil and the bed must outcrop. (Its outcrop would be the 0 m isopachyte.) Bed (or stratum) isopachytes, concerned with beds of varying thickness, are dealt with on p. 73.

SECTIONS ACROSS PUBLISHED GEOLOGICAL SURVEY MAPS

Chesterfield: 1″ Geological Survey map (Sheet No. 112)

Find the major unconformity on this map. Locate some faults that are older than the Permian beds and some that are younger. Draw a section along a line across the map ensuring that it passes through the Ashover anticline in the south-west.

Leeds 1:50,000 Map No. 70

Draw a section along the 'Line of Section' engraved on the map to show the geological structures.

6

MAP SOLUTION WITHOUT STRUCTURE CONTOURS

This chapter introduces the interpretation of geological maps without recourse to drawing structure contours. It also acts as a revision of previous chapters.

Outcrop patterns depend on the geological structures and their intersection with the topography. In general, the topography is related to the underlying geology in a number of ways: in areas of relatively simple geological structures younger beds are found forming the hills while older strata outcrop and are exposed in the valleys. Hard rocks are more resistant to erosion than softer rocks, horizontal beds (and beds with low angle of dip) tend to produce steep escarpments, whereas steeply dipping beds generally give rise to more gentle slopes. Anticlines, their beds weakened by tension, are easily eroded. The result of these relationships is that structures such as unconformities, folds and faults, etc., can often be recognized from the typical outcrop patterns they produce. In many cases, therefore, it is possible to deduce the geological structure of an area from the outcrop patterns and other given information although, for one reason or another, it is not possible to construct structure contours (strike-lines).

Particularly in Geological Survey maps, much information is given on the map and in the margins. Dip of strata is shown at many places and anticlines and synclines are usually shown, often with the axial trace indicated. Faults are shown, often with an indication of the kind of fault, normal, reverse or wrench. The direction of downthrow is usually given and in some cases the amount. In areas of metamorphic geology information on foliation and lineation is given and shear zones may be indicated. In the margins of the map there will be symbols for these features.

The geological strata and igneous rocks may be arranged by their igneous and sedimentary classification (see key to Map 23) or they may be arranged by age groups (Fig. 30(a)). However, in areas of chiefly sedimentary rocks the strata are usually presented in the form of a stratigraphical column (Fig. 30(b)). This column will show the thicknesses of beds and their age. Lithologies, fossils and other characteristics may be described. Unconformities will be shown. Where given, this kind of information used in conjunction with outcrop patterns and other map evidence is invaluable.

The simplest cases to interpret are where geological structures outcrop on a flat surface. We find this when making large-scale geological maps of intertidal wavecut platforms and other erosion surfaces. While structure contours cannot be drawn, dip can usually be measured at a number of places. Exposure may

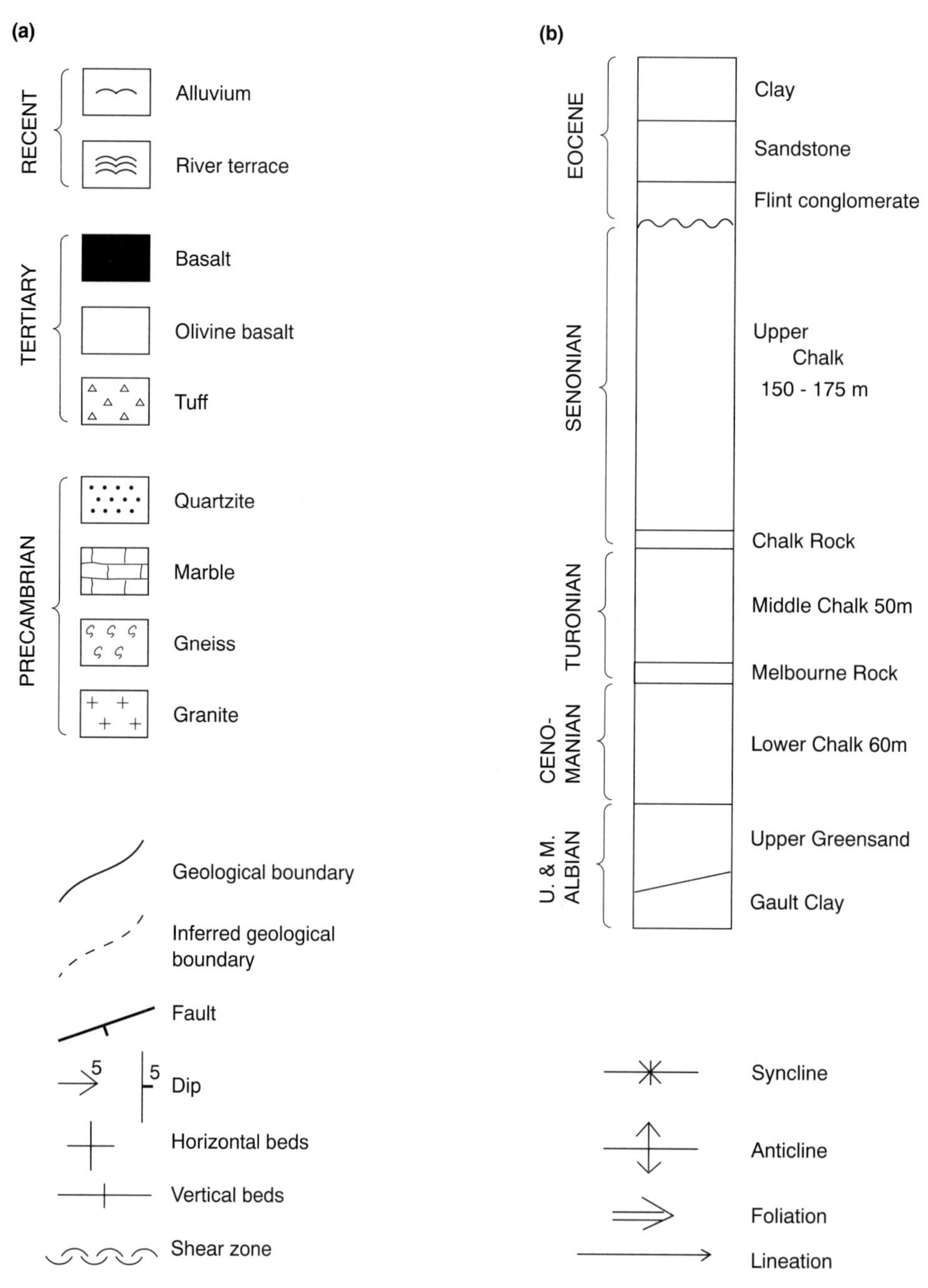

FIGURE 30 Two ways in which information may be presented on the margin of a geological map. (a) Rock types grouped together according to their age, Precambrian, Tertiary, Recent (and in this case, coincidentally grouping together similar rock types (metamorphic, igneous, sedimentary)). (b) A stratigraphic column. Below some of the more common symbols encountered on a map, such as these, may be shown on the map margin. See also the Key to Map 23.

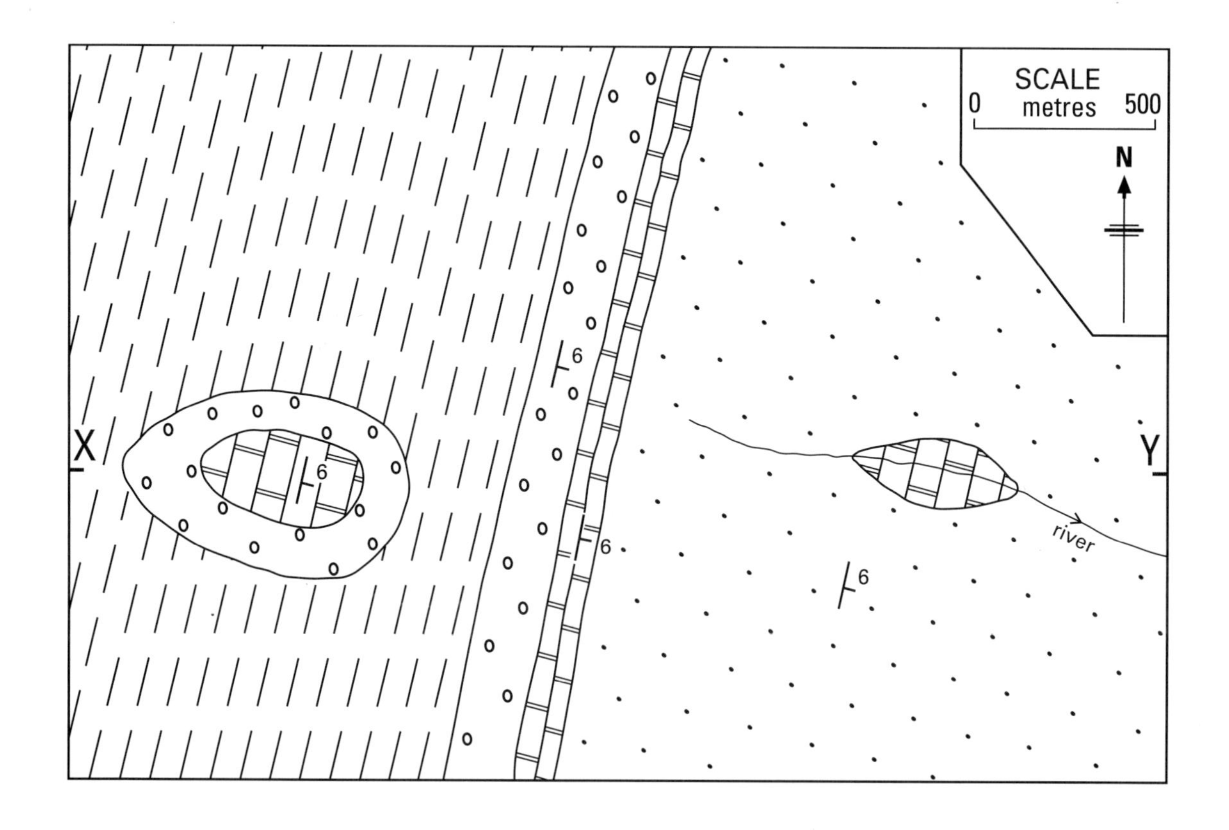

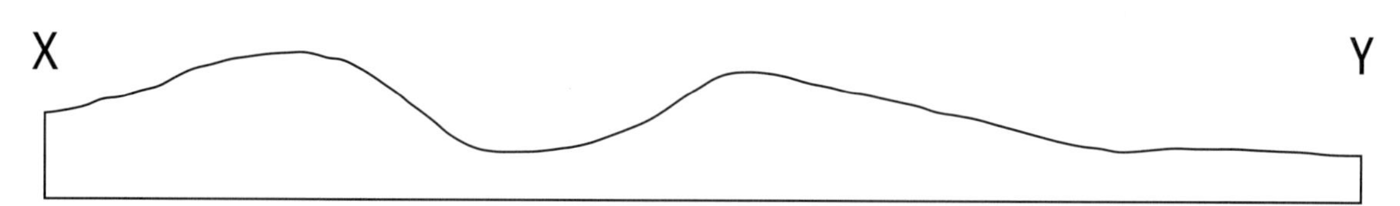

Map 19 Draw a section along the line X–Y on the topographic profile provided. Indicate on the map and on the section the inlier and the outlier. Reproduced from Fig. 14 of *Simple Geological Mapwork* by W.E. Johnson (London, Edward Arnold, 1976) with kind permission of the author.

be almost 100%, if not obscured by loose blocks or other recent deposits (see Map 22).

At the other end of the scale broad plains or plateaux may be nearly flat and horizontal for tens of kilometres, even 100 km or more. Although they may be hundreds of metres above sea level (for example in West Texas or Arizona) it may be impossible to obtain the data needed to construct structure contours. The geological structures present must be deduced from outcrop patterns and information given in the legend of the map.

Typically, on maps drawn on a scale of 1:50,000 structure contours are seldom straight or parallel over the distances being covered. There is seldom sufficient evidence to permit structure contours to be drawn, although dip and strike will be given at a number of points where they have been measured and recorded by the field geologist. Additional useful data

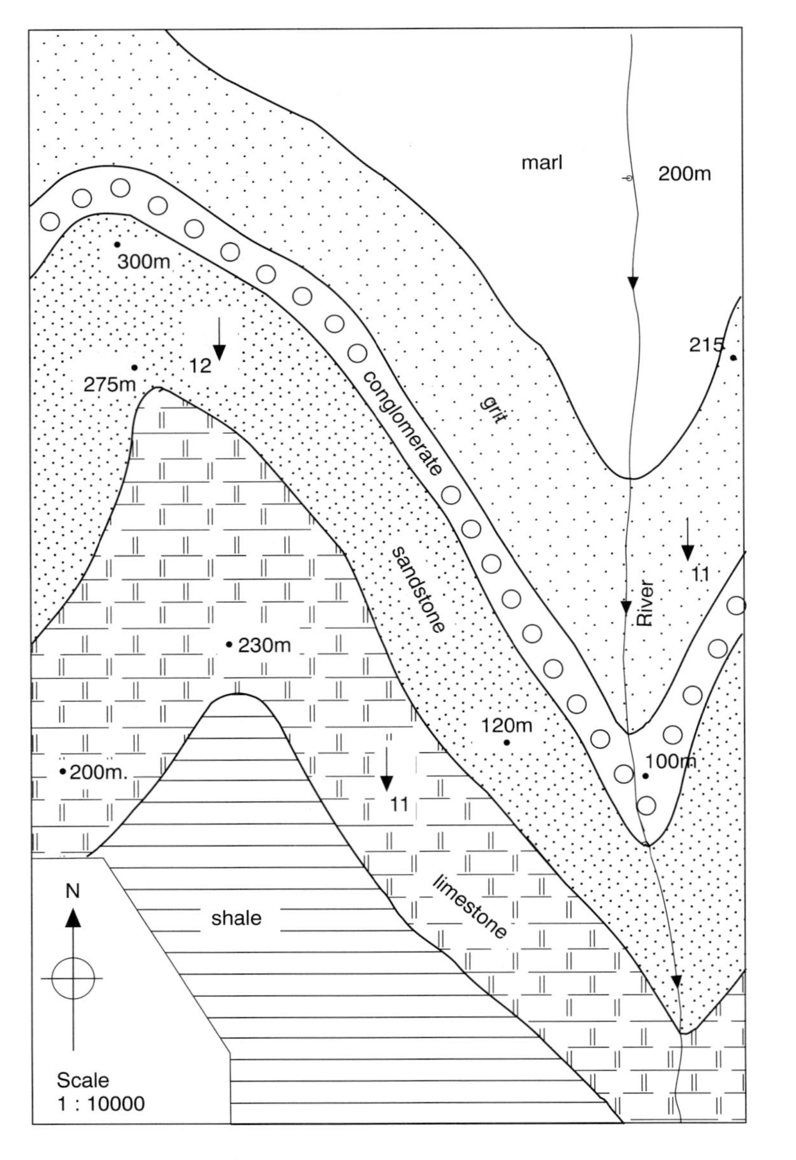

FIGURE 31

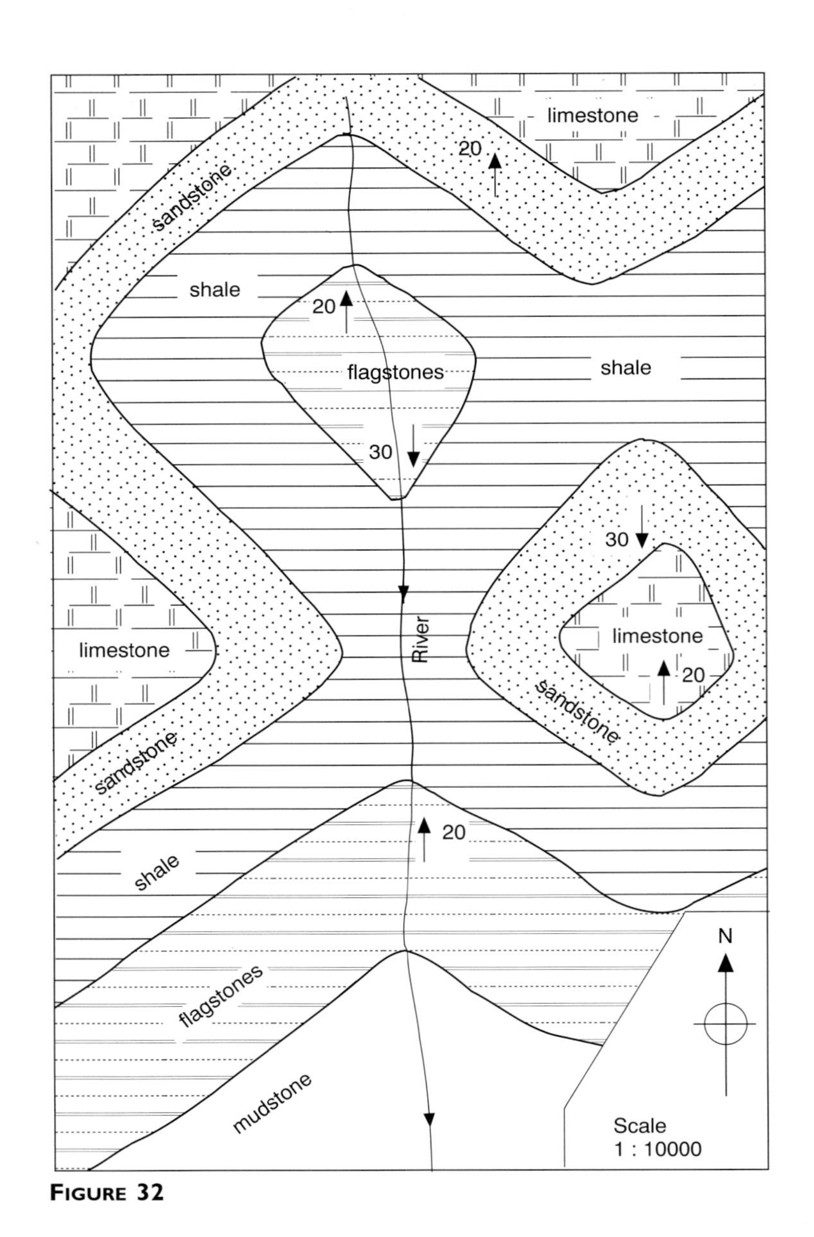

FIGURE 32

may be deduced from outcrop/contour intersections but, in general, on maps of this scale the interpretation is largely dependent on the outcrop patterns observed, together with a wealth of data given in the margins of the map (the legend). Problem Maps 20 and 21 are drawn on this scale.

(A warning must be given that on published survey maps there sometimes appears to be a dip given which does not appear to fit in with the structural pattern deduced from the outcrop pattern and other information. Such data is not discarded since it is not clear whether this may be a local variation of significance.)

To recapitulate, consider outcrops on a near-horizontal plane surface. We saw in Chapter 1 that older beds dip under younger beds. This is a consequence of beds being deposited one on top of another, in any sequence the oldest beds at the bottom, the youngest at the top (known by the rather impressive name of the Law of the Order of Superposition). Following deposition, tilting and subsequent erosion produce this effect.

We also saw in that chapter (p. 00) that the steeper the dip, the narrower the outcrop of any bed. The extreme case occurs when beds are vertical – dipping at 90° – and the width of outcrop equals the actual thickness of the bed.

Where sequences of outcrops of beds are repeated in the reverse order, folding is usually responsible. The Order of Superposition (the relative ages of the strata) will indicate the direction of dip in the limb of a fold. It can readily be seen whether a fold is an anticline or a syncline and whether overfolding is present (see Maps 11 and 13). The repetition of sequences of outcrops, or the suppression of parts of a sequence of outcrops is caused by faulting. This was dealt with in the last chapter. Unconformities are indicated in the field by erosion surfaces and such information may be given on a geological map. However, the essential evidence on the map is the difference in the direction of strike in the pre-unconformity sequence of strata and the direction of strike in the post-unconformity strata. (Only very rarely will they have a similar strike direction by chance.) Generally an unconformable bed lies on several beds of the older series since they may have been tilted or folded during the period of non-deposition.

Map 19 illustrates a typical outcrop pattern of dipping strata giving rise to an escarpment with a steep scarp slope and gentle dip slope. Imagine for a moment that there were no dip arrows on the map but that we knew the stratigraphical sequence – sandstone, limestone, conglomerate, marl. The oldest bed, the marl, outcrops in the west of the area while the youngest bed, the sandstone, outcrops in the east. Erosional features have produced an inlier and an outlier.

Figure 31 is a simple map which shows beds outcropping in an area of simple topography: a north–south river valley with a high ridge to the west is confirmed by a number of spot heights. Such information on the geology as is available shows an almost uniform dip of 11° to the south. The beds dip downstream (and are steeper than the slope of the valley floor), therefore their intersection with the topographic surface gives rise to V-shaped outcrops in the valley, that 'V' downstream. Naturally, the outcrops bend the other way over the ridge. Compare the outcrop pattern here with those on Maps 4 and 10.

Figure 32 is a map of a north–south valley sloping southwards and flanked either side by hills. Where beds dip upstream the outcrops 'V' upstream. Where beds dip downstream the beds 'V' downstream. This gives rise to lozenge- or 'diamond'-shaped outcrops, characteristic of folded strata where the fold axes tend to be more or less at right angles to the drainage. See, for example, Maps 12 and 18.

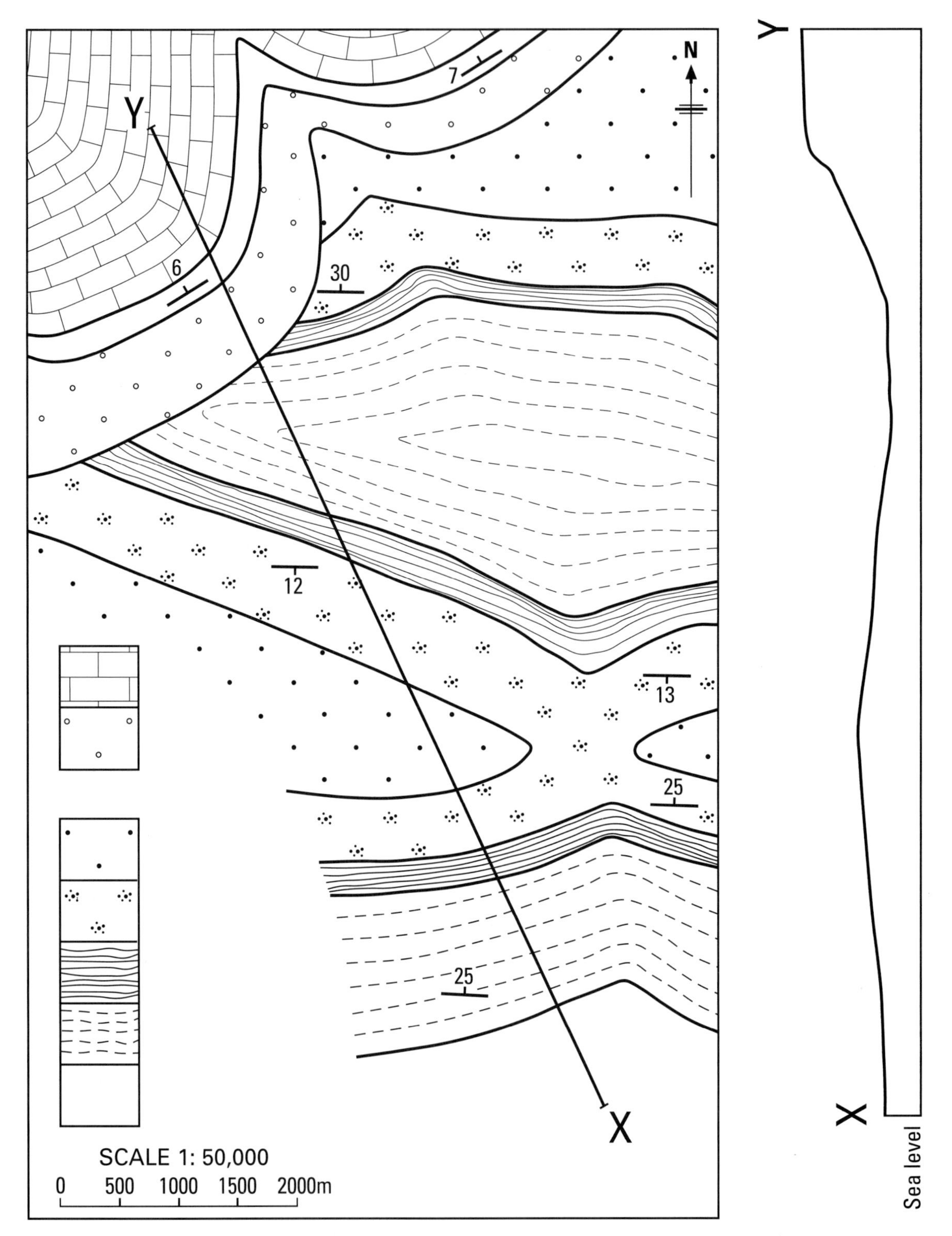

Map 20 Describe the structures shown on the map. Draw a section along the line X–Y on the profile provided. You will note that relief is slight except for a prominent scarp feature in the north-west of the area. Before commencing, check what the apparent dips will be in the direction of the line of the section, consulting the table on p. 117.

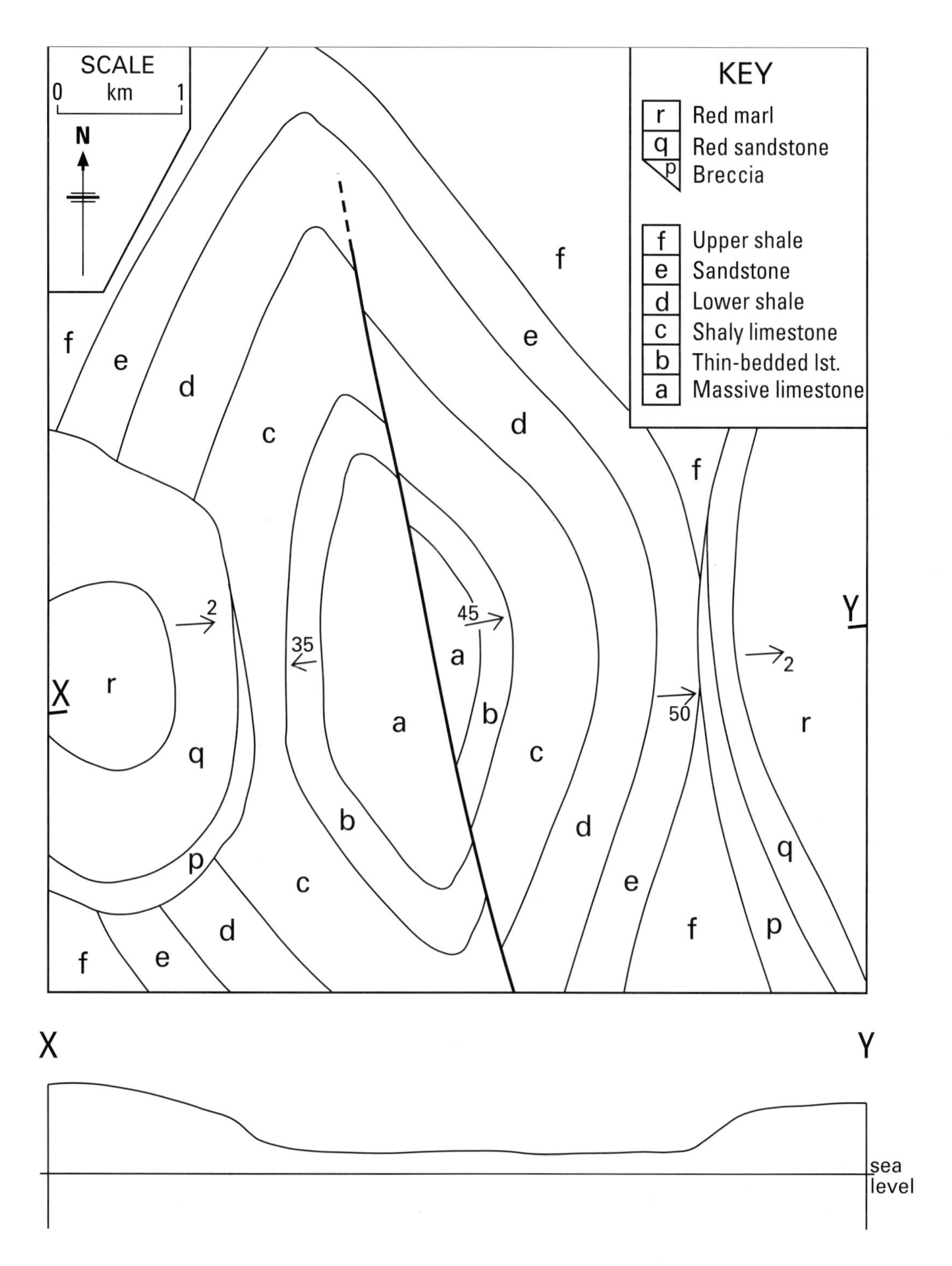

MAP 21 Describe the geology of the area of the 1:50,000 scale map. Draw a section along the line X–Y on the profile provided. Show the structures to a depth of 250 m or more below sea level (which from a structural geology viewpoint is a quite arbitrary datum).

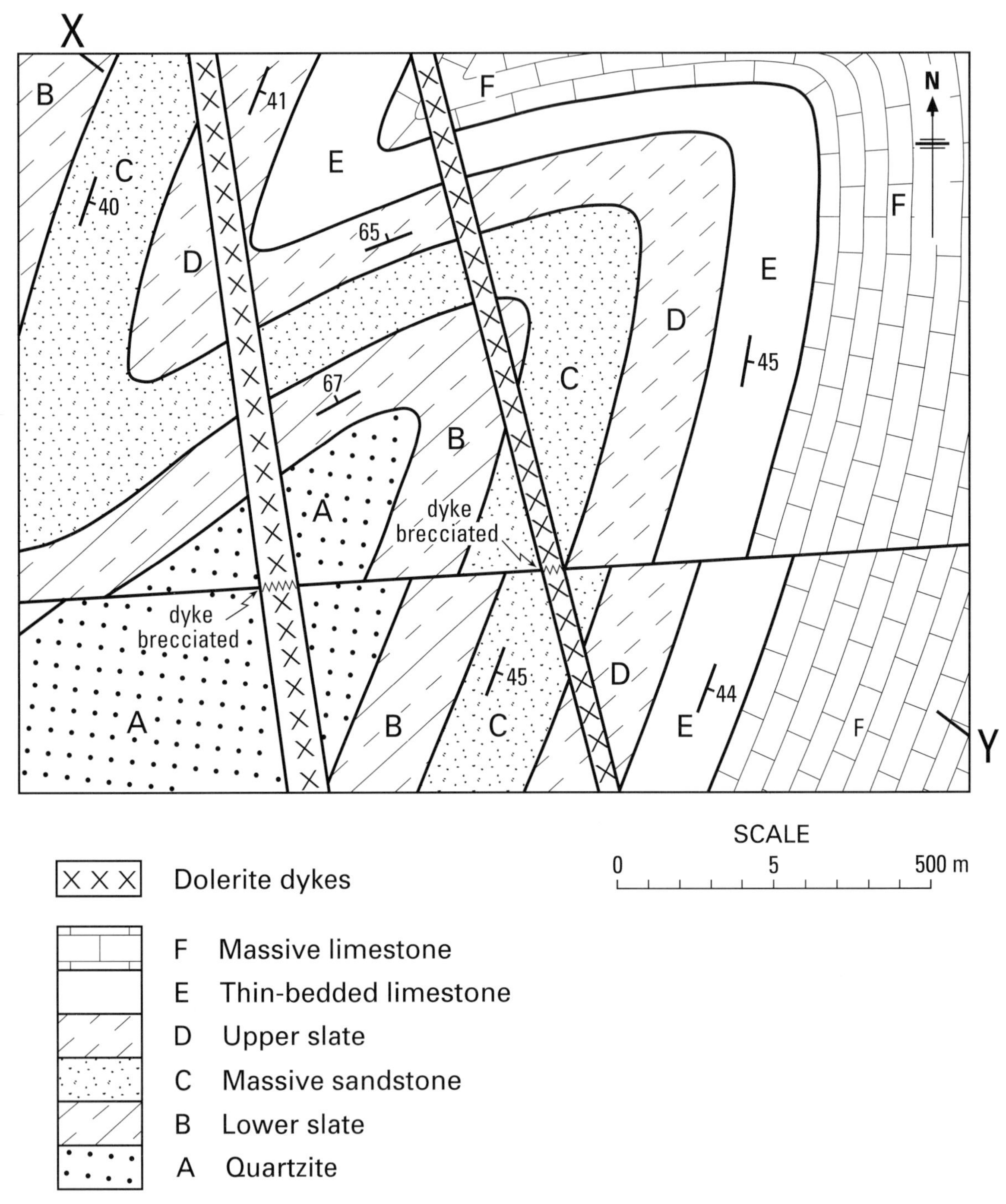

Map 22 The map shows the geological exposures on a small area of wavecut platform. The surface is nearly horizontal and almost flat, exposure is very good (except for seaweed and fallen blocks from the adjacent cliffs, which have been omitted to simplify the map). Draw a section along the line X–Y to show the structures present, assuming a flat ground surface. Use the same vertical scale as the horizontal scale, i.e. no vertical exaggeration, so that angles of dip will be correctly represented.

Note on Map 21. Note the asymmetry of the fold, which is a pericline. Note that the map includes an example of overlap as well as overstep. The throw of the fault decreases northwards and it dies out within the outcrop of bed d. Of course, no faults go on for ever but faults of variable throw introduce problems usually reserved for advanced textbooks.

Note to Map 22. There are some structural elements in Maps 22 and 23 not yet dealt with. You may be able to deduce the structures present but, at least in the case of Map 23, you are strongly advised to read Chapter 8 on igneous rocks before attempting to answer the questions on this map.

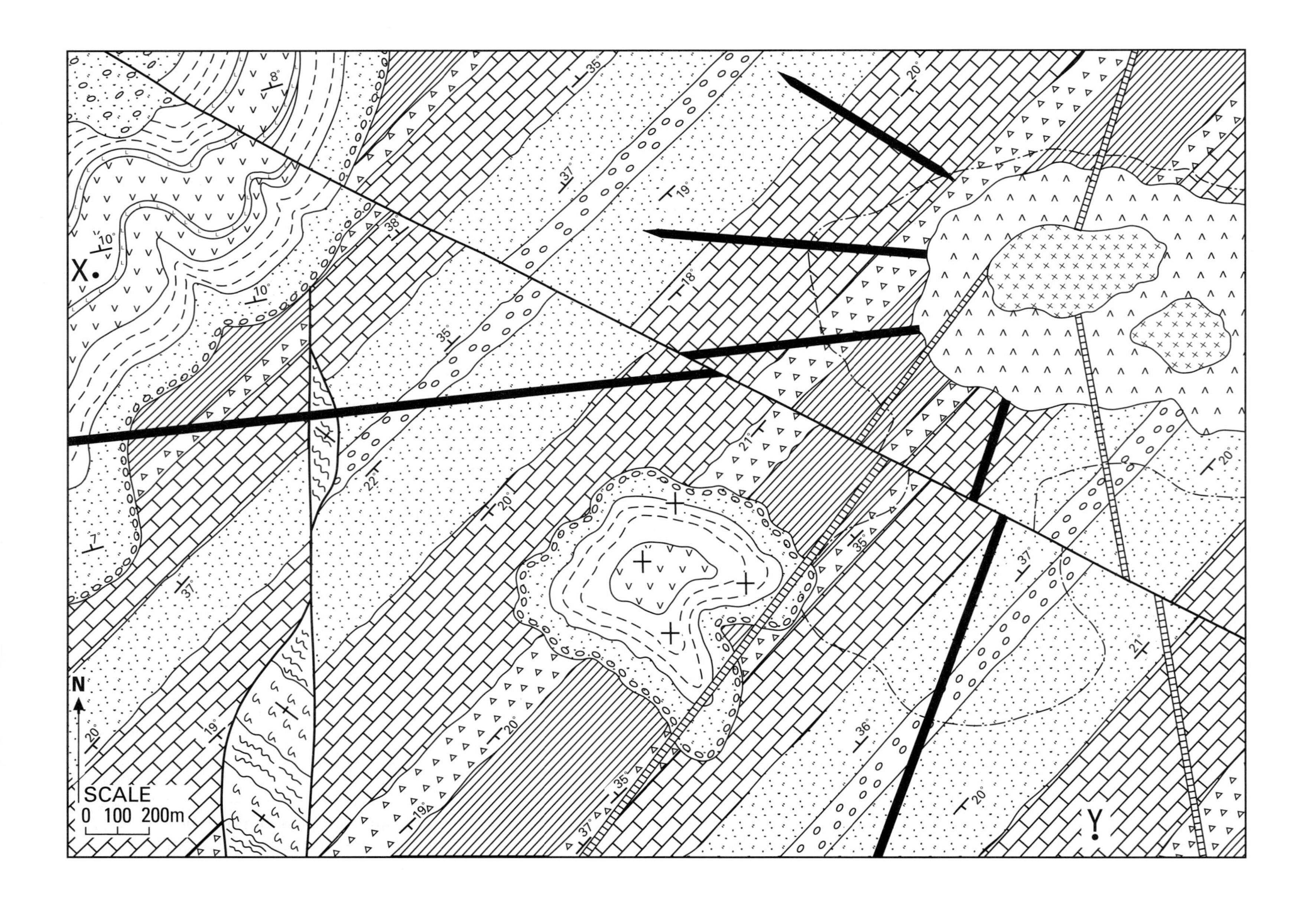

X
Y
N
SCALE
0 100 200m

Map 23 There are two faults: one runs north–south and the other runs north–west–south–east. Determine, from the outcrop patterns, whether the displacement has been vertical or horizontal in each case. Are the folds symmetrical or asymmetrical? What is the evidence? Mark on the map the major unconformity and indicate inliers and an outlier. Can you decide in which direction the lower lava flowed? Which side of the plutonic intrusion has the steepest side? What surface evidence gives you a clue? Indicate on the map where a still buried pluton may lie.

Describe the geology of the area of the map. Draw a section along the line X–Y on the profile provided. The geological events that occurred in the area can be relatively dated and then arranged in chronological order, providing a summary of the geological history. For further guidance on this refer to 'Description of a Geological Map' on p. 109.

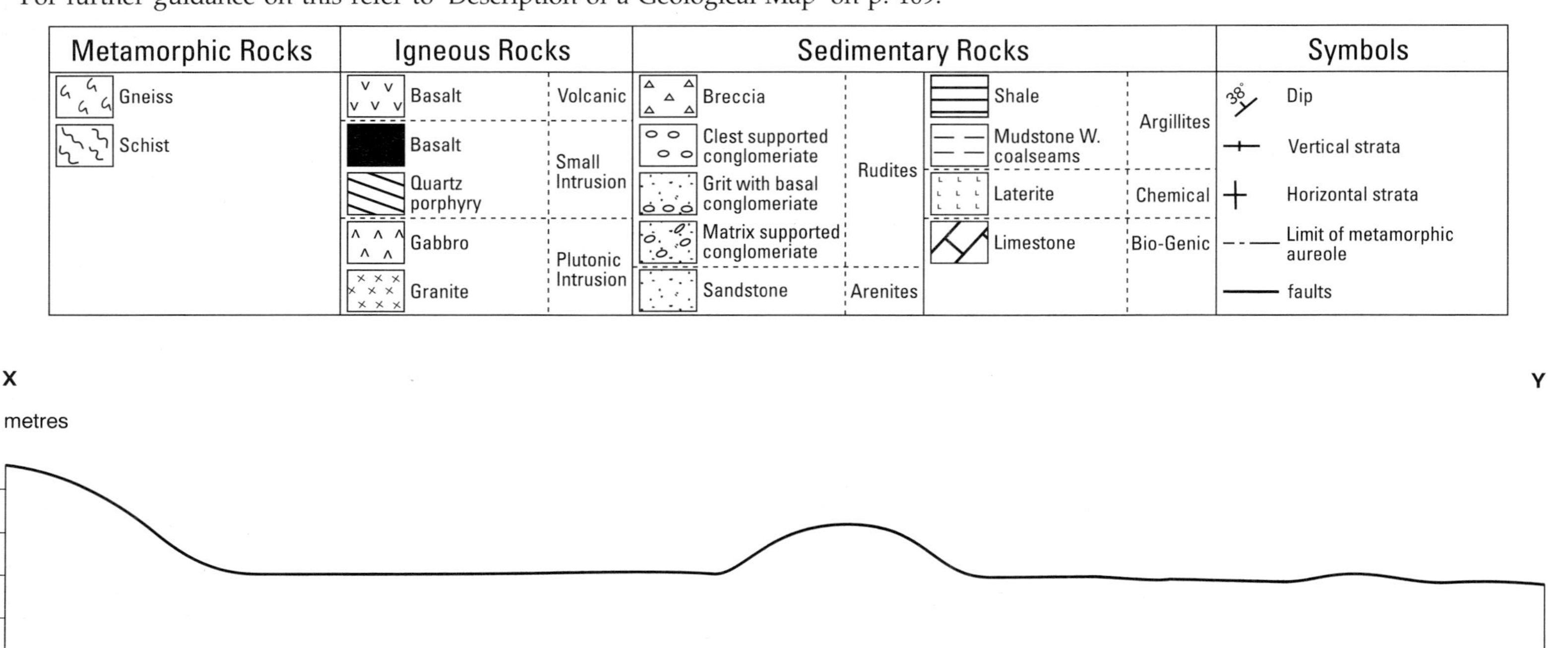

Topographic profile of section X–Y.

7

MORE FOLDS AND FAULTED FOLDS

PLUNGING FOLDS

In the folds studied so far, the axis of the fold has been horizontal. The axis is the intersection of the axial plane with any bedding plane. (Make a synclinal fold by taking a piece of paper and folding it in two to make a simple V-shape. Drop a pencil into this and it will assume the position of the axis of the fold.) Such a fold, with the axis horizontal, is called a *non-plunging* fold. The outcrops of the limbs tend to be parallel (but of course are affected by the configuration of the ground) since the structure contours drawn on one limb are parallel to those drawn on the other limb. (They have, as well, been parallel to the axial plane and, in cases where the axial plane was vertical, parallel to the axial trace (= outcrop of the axial plane).)

Where the fold axis is not horizontal, but is inclined, the fold is described as plunging. Some fold structures can be traced for many kilometres, others are of much more limited extent. Eventually a fold will die out by plunging (Fig. 33). Anticlinal folds which die out by plunging at both ends, giving rise to elliptical outcrop patterns, are called periclines and are a common occurrence.

Simple folds which have been subsequently tilted by further earth movements will also plunge. This is, furthermore, the simplest way

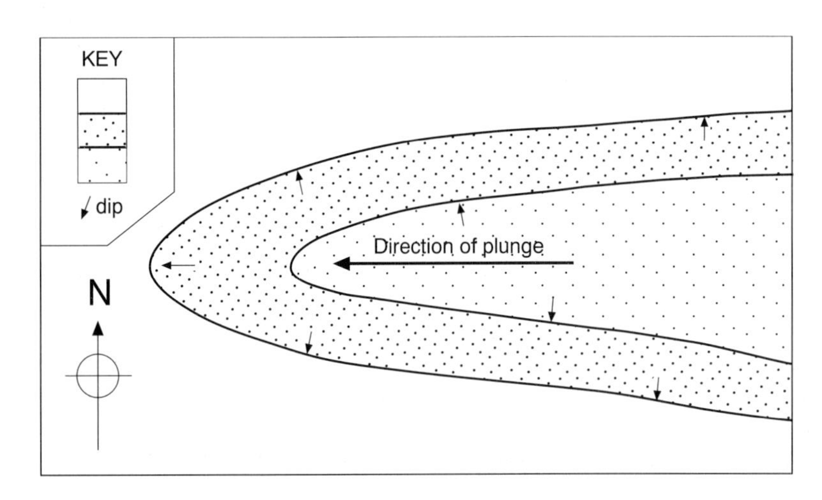

FIGURE 33 Map showing the outcrops of anticlinally folded beds, plunging to the west.

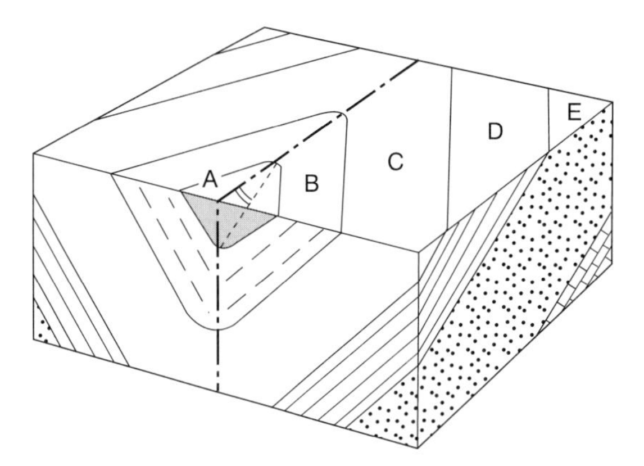

FIGURE 34 Block diagram of a plunging syncline.

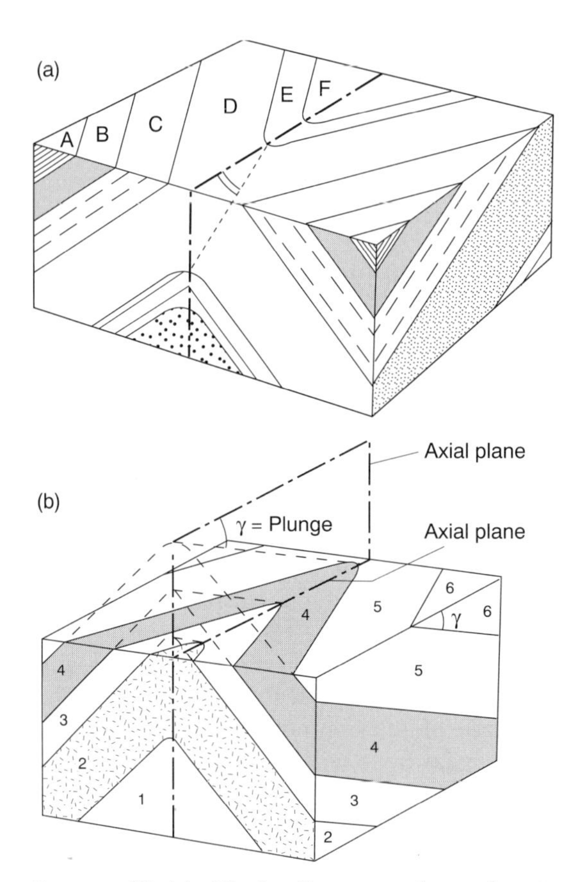

FIGURE 35 (a) Block diagram of a plunging anticline with the plunge towards the observer. (b) Plunging anticline with the plunge away from the observer, eroded strata shown in broken lines.

of considering such structures although they may originate in other ways. This type of fold is referred to in some earlier books as a 'pitching' fold, and not infrequently the terms 'pitch' and 'plunge' are used synonymously. The advanced student should consult F.C. Phillips's *The Use of Stereographical Projection in Structural Geology*, 3rd edition (London, Edward Arnold, 1971), which deals fully with these terms. It is best to restrict the term 'pitch' to such linear features as striae, slickensides or lineations on a plane. They have both plunge (the angle they make with the horizontal) and pitch (the angle they make with the strike direction of the plane). Reject the use of the term pitch when referring to folds. Since problems entailing pitch can be solved (quite simply) by the use of stereographic projection, they are not included in this book.

The effects of erosion on plunging folds are seen in Figs 34 and 35. The outcrops of the geological interfaces of the two limbs of a fold are not parallel. While the structure contours are parallel for the beds of each limb, those of the two limbs converge, meeting at the axial plane (Fig. 36).

Calculation of the amount of plunge

Just as the inclination of the beds (dip) can be calculated from the spacing of the structure contours (see p. 8) measured in the direction of dip, so can the plunge of the fold be calculated from the spacing of the structure contours measured in the direction of plunge, i.e. along the axis. The plunge of the fold shown in Fig. 36 is, expressed as a gradient, 1 in 3.5 (to the north), if the scale of the diagram is 1 cm = 200 m, since the structure contour spacing measured along the axis is 1.75 cm. (The dip of the bedding plane of each limb is 1 in 2: check that this is so.)

Where we have no structure contours enabling accurate calculation of the angle of plunge, it is possible to give some estimate of it from the width of outcrop. In Figure 33 the outcrop of the coarsely stippled bed is about

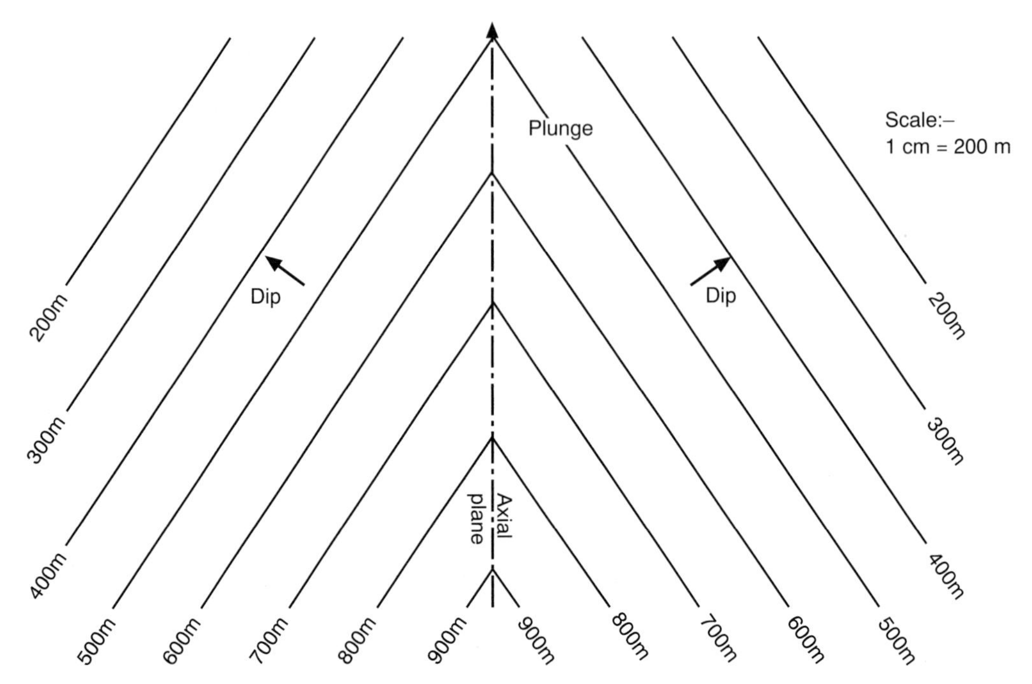

FIGURE 36 Map of the structure contour pattern of a plunging fold.

twice as wide in the direction of plunge as it is in the limbs of the fold. We can conclude that the amount of plunge is considerably less than the angle of dip of the limbs of the fold. The relationship of dip to width of outcrop was discussed in Chapter 1.

THE EFFECTS OF FAULTING ON FOLD STRUCTURES

We have seen in Chapter 5 that the effect of dip–slip faulting (normal and reversed faults), followed by erosion, produces a lateral shift of outcrop in the direction of the dip of the strata on the upthrow side. Observe that on Figure 37 the vertical axial plane of a symmetrical fold is not displaced but the dipping planes show a shift of outcrop. Remember that the lower the angle of dip the greater the shift of outcrop.

Since the shift of outcrop is down-dip as a result of erosion, it follows that on the upthrow side of a fault the outcrops of the limbs of an anticline are more widely separated (than on the downthrow side). Conversely, the outcrops of the limbs of a syncline are closer together on the upthrow side (than on the downthrow) (Fig. 37).

On an asymmetrical fold, there will be a more pronounced shift of the outcrops of the limb with the lower angle of dip and a less pronounced shift of the outcrops of the steeper dipping limb. In the 'limiting' case of a monoclinal fold with a vertical limb there will be no shift of outcrops of the vertical beds. The axial plane of an asymmetrical fold dips, therefore the axial trace will be displaced.

In the case of overfolds – where both limbs dip in the same direction (see Fig. 16) – naturally the outcrops of both limbs will be shifted in the same direction by a dip–slip fault. Of course the outcrops of the shallow limb will be shifted more than those of the steeper dipping overturned limb.

Displacement of folds by strike–slip (wrench) faults

A wrench fault causes a lateral (horizontal) dislocation or displacement of the strata which may in some case be of many kilometres. The effect of a wrench fault is to displace the outcrops of beds, always in the same direction – and its effect on the outcrops of simply dipping strata in certain circumstances is similar to that of a normal fault: it may not be possible to recognize from the evidence a map can provide whether a fault has a vertical or a lateral displacement (see p. 46). However, the effects of a wrench fault on folded beds is immediately distinguishable from the effects of a normal fault since it will displace the outcrops of both limbs by an equal amount and in the same direction (whether the fold be symmetrical or asymmetrical) and, further, the outcrops of a bed occurring in the limbs of a fold will be the same distance apart on both sides of the fault plane (Fig. 38). The axial plane will, of course, be laterally displaced by the same amount as the outcrops of the beds occurring in the limbs of the fold, and this displacement occurs whether the axial plane is vertical (in a symmetrical fold) or inclined (in an asymmetrical fold). Compare Figs 37 and 38.

Calculation of strike–slip displacement

Where any vertical phenomenon is present, for example the axial plane of a symmetrical fold, vertical strata, igneous dyke, etc., the lateral displacement by the fault can be measured directly from the map. If the strata are dipping and the fault is a purely strike–slip fault with no vertical component, or throw (i.e. it is not an oblique-slip fault, see Fig. 25), then we can simply find the lateral shift by measuring the displacement of any chosen structure contour on a given geological boundary. Where strata are folded it is easy to measure the lateral displacement of fold axes or axial planes (Fig. 38).

Faults parallel to the limbs of a fold

So far we have considered faults that were perpendicular to the axial planes of the folds or, at least, cut across the axial plane. A fault which is parallel to the strike of the beds forming the limb of a fold is, of course, a strike

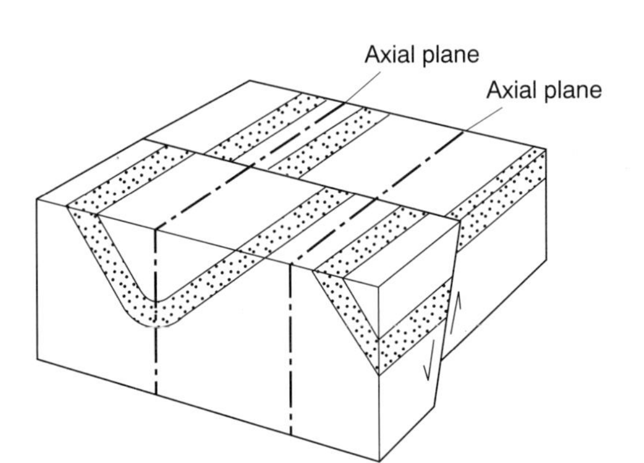

FIGURE 37 Block diagram showing folded strata (one bed is stippled for clarity) displaced by a normal fault.

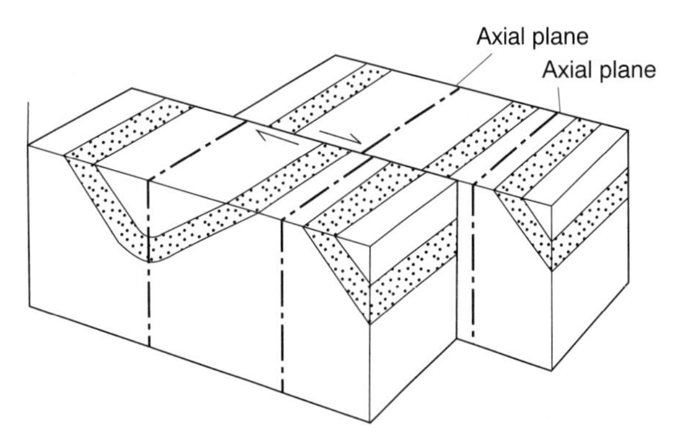

FIGURE 38 Block diagram showing similar beds to Fig. 37 displaced by a wrench fault.

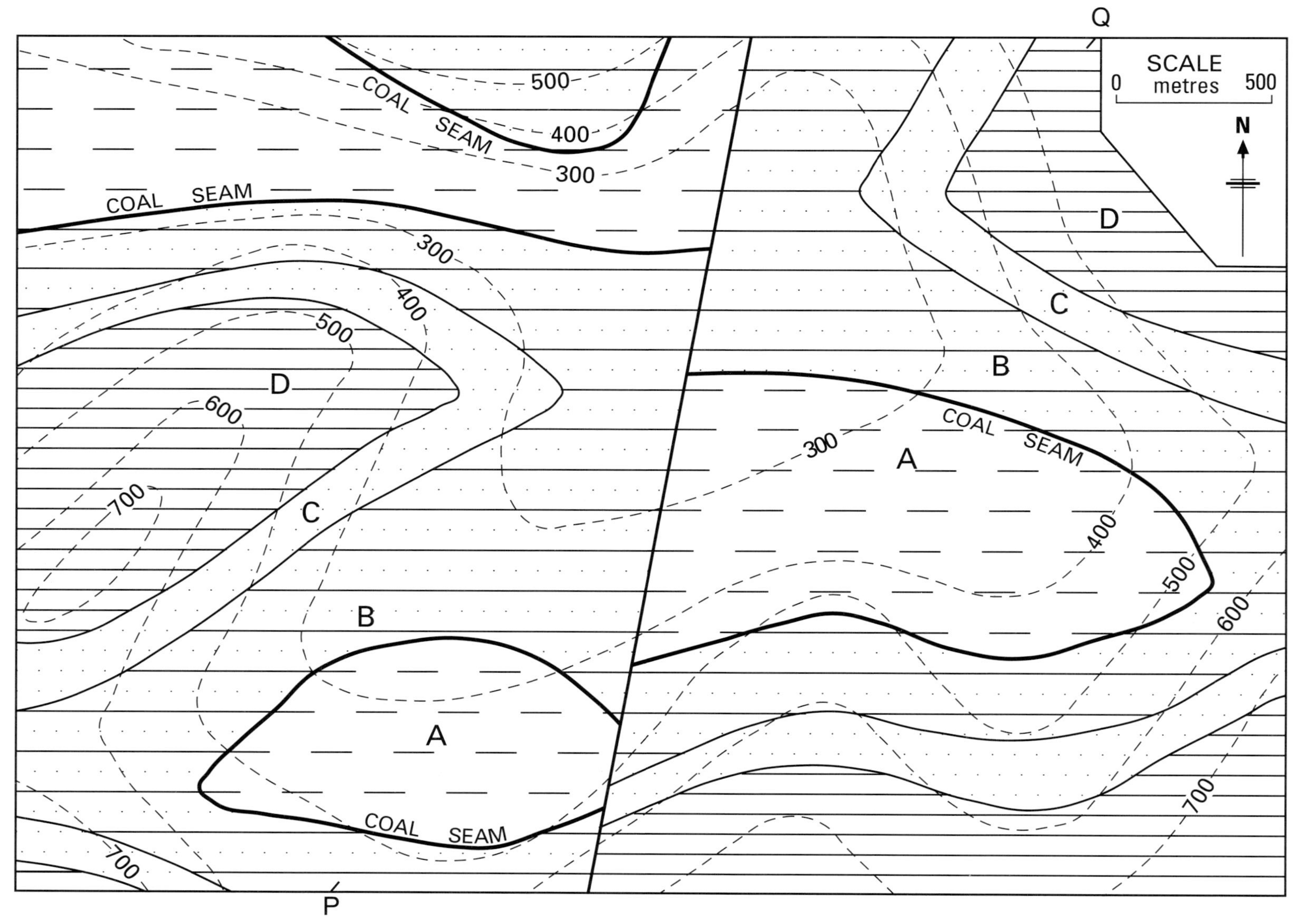

MAP 24 What kind of fault occurs on this map? Has it any vertical displacement? Draw a section across the map along the line P–Q intersecting the fault. Indicate on the map anticlinal and synclinal axial traces.

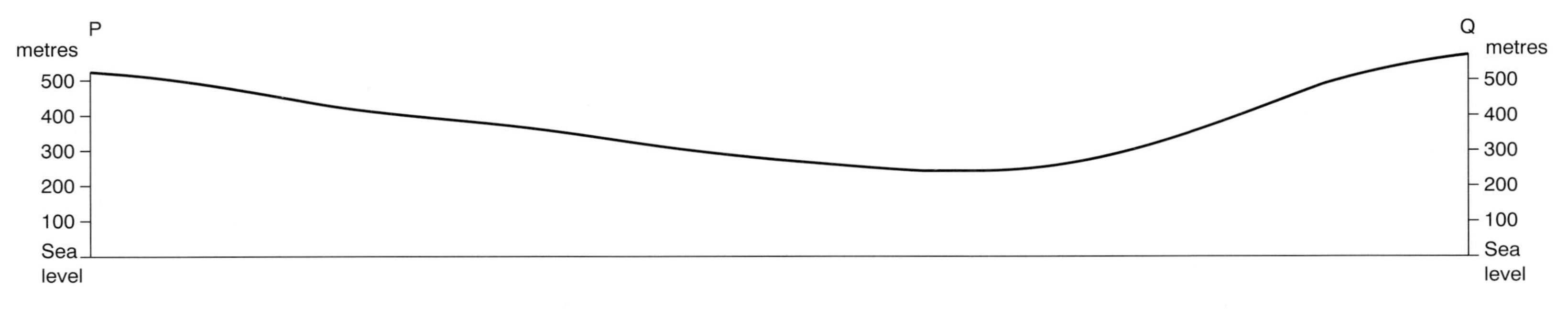

Topographic profile of section P–Q.

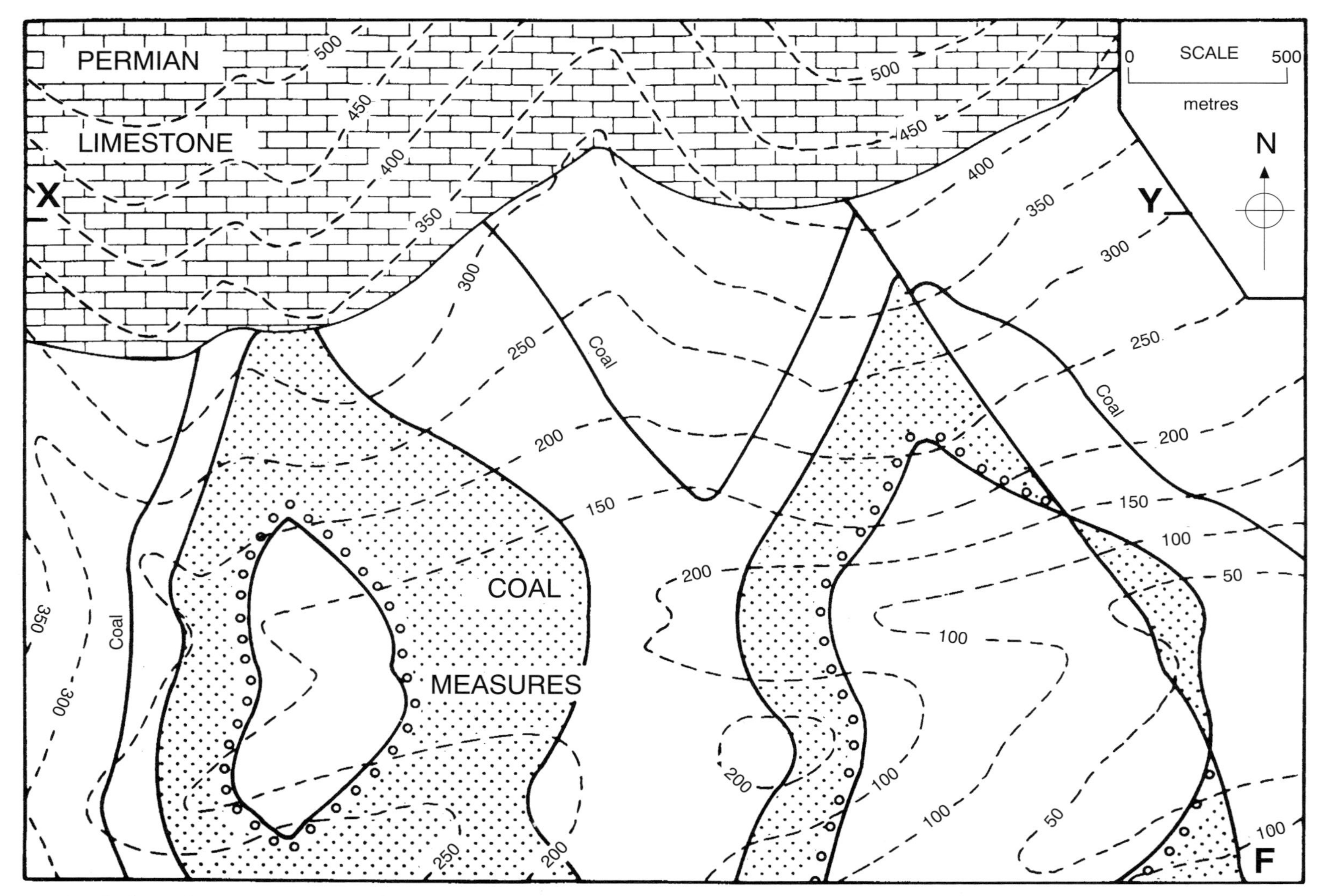

Map 25 Deduce the main structural features from the outcrop patterns. Insert the fold axial traces on the map. Draw a geological section along the line X–Y. Insert on the map the sub-Permian Limestone outcrops of the coal-seam, the sandstone and the fault plane. To construct the latter it will be necessary to draw structure contours for the fault plane in order to deduce its dip.

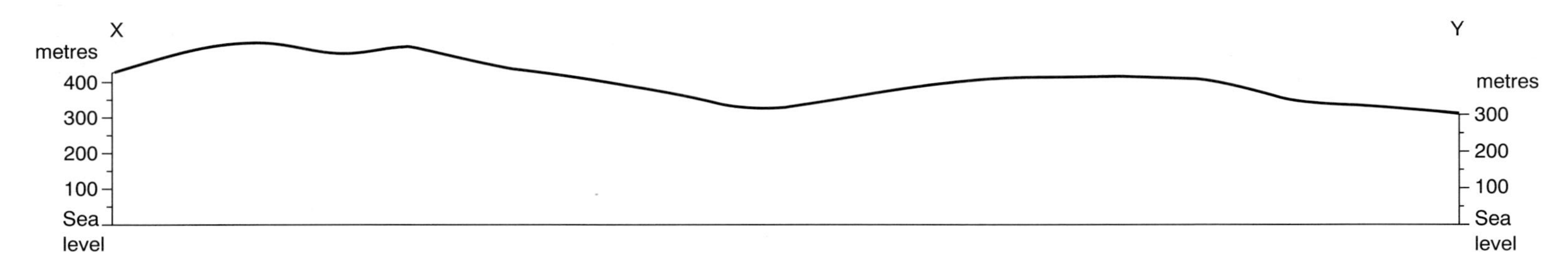

Topographic profile of section X-Y.

fault. This will cause either repetition of the outcrops or suppression of the outcrops of part of the succession, in the manner discussed on p. 42.

SUB-SURFACE STRUCTURES

An interesting and important geological consideration is the deduction of the disposition of strata beneath an unconformity. In effect we are considering the 'outcrop' pattern on the plane of unconformity. If we could use a bulldozer to remove all strata above that plane we would see the earlier, pre-unconformity strata outcropping.

This problem was touched on briefly in Map 10 where unconformity was introduced. If not already completed, return to Map 10 and insert the sub-unconformity outcrop of the coal-seam. (Remember that topographic contours are now irrelevant; the surface we are considering is the plane of unconformity which is defined by the structure contours drawn on the base of Bed Y. Since both sets of structure contours are in each case straight, parallel and equally spaced – representing constant dips – their intersections giving the sub-Y outcrop of the coal-seam should lie on a straight line.)

POSTHUMOUS FOLDING

After the strata in an area have been laid down they may be uplifted, folded and eroded as we have seen in the last chapter. Further subsidence may cause the deposition of strata lying unconformably upon existing beds. Later still the processes of uplift and folding may recur, when this second period of folding is said to be posthumous – if the fold axis in the younger beds is parallel to, and approximately coincident with, the similar fold axis in the older beds. The trend

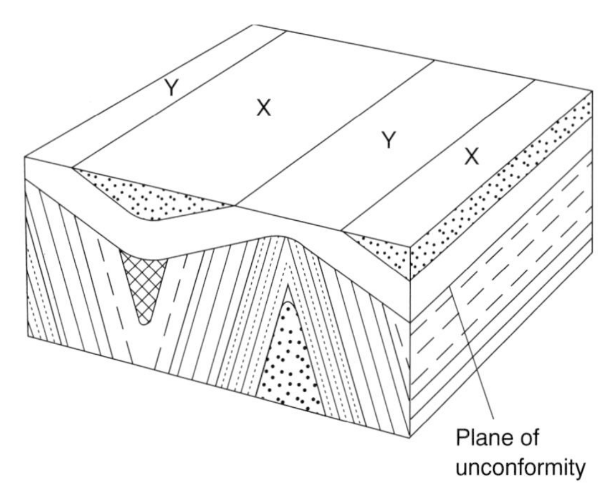

FIGURE 39 Block diagram showing the effects of posthumous folding.

of the two sets of folds may be parallel and the fold axes may coincide, as in Fig. 39, or they may be parallel but not coincident. An excellent example of this can be seen demonstrated by the Upper and Lower Carboniferous age beds in the Forest of Dean area included on the BGS 1:50,000 map of Monmouth (Sheet 233).

Since the folding is of two ages, the trends of the two sets of folds may be in quite different directions and it is then called superposed folding or cross-folding and is not included in posthumous folding; however, there is usually a tendency for the earlier folding to exercise a 'control' over the later folding so that, more commonly, the trends are parallel or subparallel.

The strata of Pembrokeshire (see the Haverfordwest and Pembroke 1″ Geological Survey maps (Sheets 227, 228)) were affected by Caledonian folding and later by Variscan (Hercynian) folding. In this area the trend of the folding of both periods is nearly east–west. By contrast, the rocks of the Lake District were gently folded with an east–west strike in Ordovician times and later by the Caledonian orogeny (phase in the mountain-folding process) with a north–east–south–west strike, the later folding being of such intensity that it

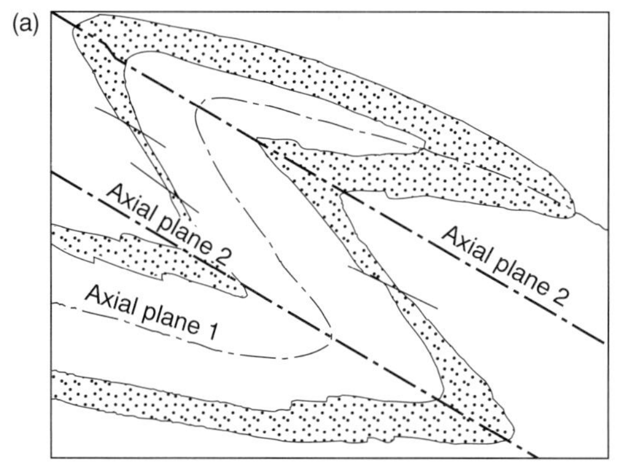

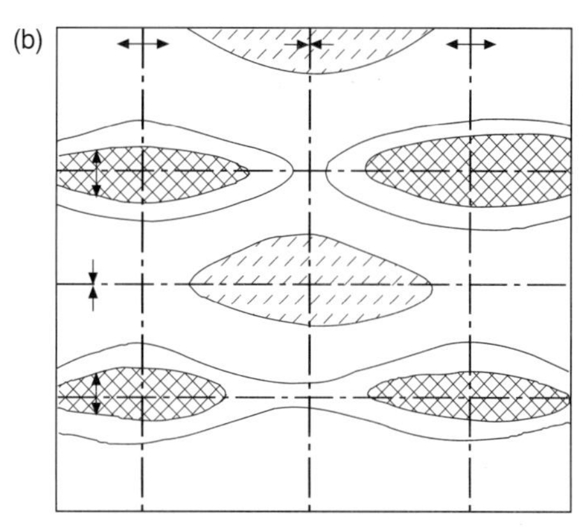

FIGURE 40 Outcrops of refolded folds on flat ground. (a) A refolded isoclinal fold, (b) second fold axes crossing first fold axes at right angles.

seems to have been independent of the control of the earlier folds.

Where the folds are of different trend it is possible to study the effects of one age of folding by taking a section parallel to the strike of the other folds. This was very neatly shown in the block diagram of a part of the southern Lake District in earlier editions of the 'Northern England' British Regional Geology.

POLYPHASE FOLDING

Strata that have already been folded may at a later time be refolded. The stress causing the refolding may be due to a later phase of the same orogeny or even due to a later orogeny. The stress of the later phase of folding may be quite unrelated to the stress direction of the earlier folding (the 'first folds'). As a result of this the trends of the two ages of folding may be quite different.

Complex outcrop patterns are produced by the interference of two (or more) phases of folding. Simple examples are illustrated (Fig. 40). In some such simple cases the early folds can be seen to have been refolded since the axial planes of the first folds are themselves folded.

BED ISOPACHYTES

All problems so far have dealt with beds of uniform thickness. However, traced laterally over some distance, strata may be seen to vary in thickness, a function of their mode of deposition. Such variations tend to be gradual and reasonably uniform within the area of a geological map sheet. Variations in the thickness of a bed are usually deduced from borehole data but may be discovered by measuring sections at geological outcrops and are occasionally revealed by variations in width of outcrop on a map.

The way in which a bed varies laterally in thickness can best be shown by constructing a series of bed isopachytes, lines joining points where the bed is known to be of the same thickness. To obtain the maximum number of control points at which the thickness of a bed can be determined, it is usually necessary to construct two sets of structure contours: those for the top of the bed and those for its base. Their intersections give the thickness of the bed. Due to the nature of the sedimentary phenomena which produce beds of varying thickness, the isopachytes will tend to be reasonably straight (or gently curving) and approximately evenly spaced.

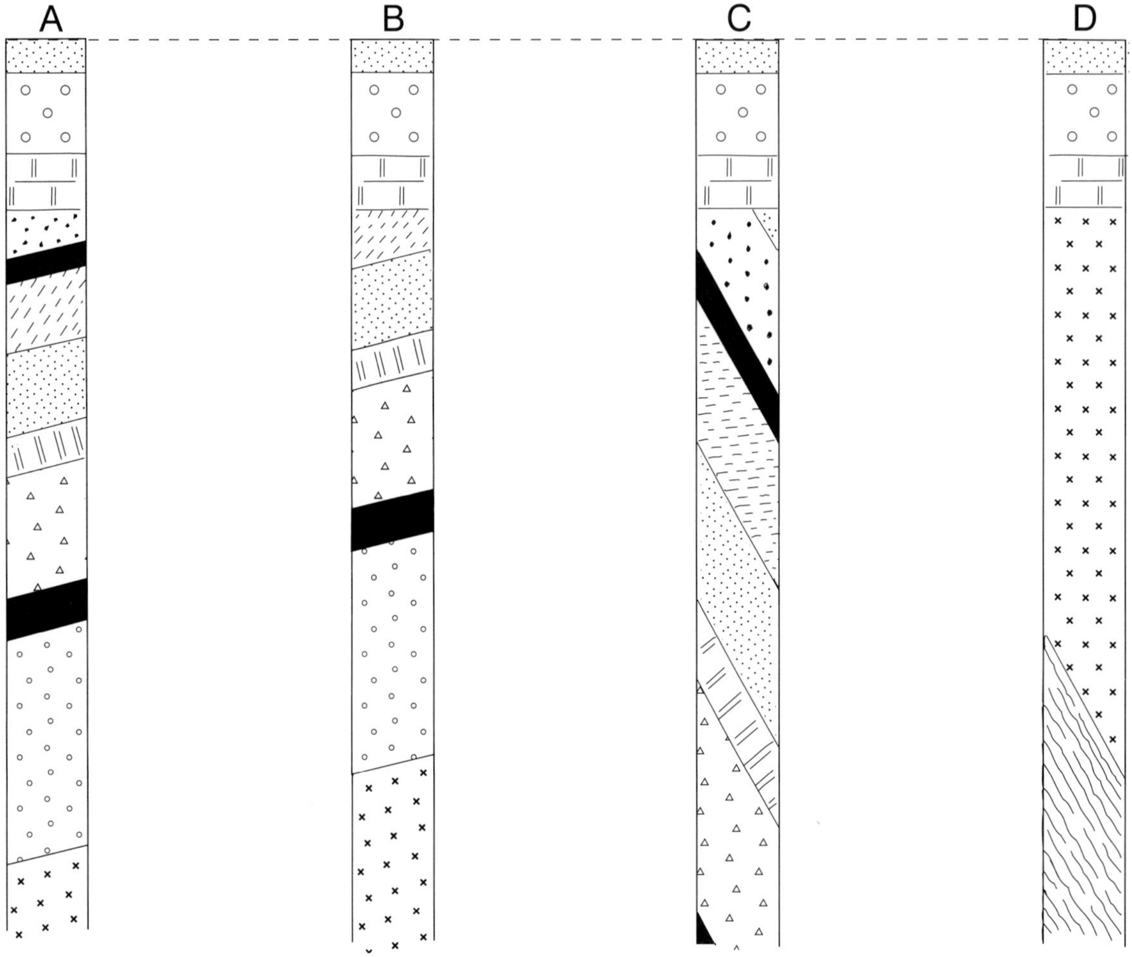

A B C D

Problem I

Four boreholes were drilled at 500 m intervals along a straight line. The vertical sections reveal in each case a succession of sedimentary rocks. Assuming that similarity of lithology and thickness of a bed indicates a probable correlation, show on the diagram the geological structures which may be present to explain the relationship of the strata in one borehole to the strata in the adjacent boreholes.

Exercise on Geological Survey Maps

1. **Haverfordwest: 1:50,000 Map No. 228**

Draw a section along the north–south grid line 197. Note the difference in amplitude of the folding in the Lower and Upper Palaeozoic rocks. Tabulate the geological events in chronological order.

Note on Problem I The higher strata in all four boreholes are the same in sequence and in having no dip – they are horizontal. Below the limestone it is clear that the beds are folded, proving the presence of an unconformity. In Boreholes A, B and C the relationship can be simply shown by sketching in the appropriate fold structure. In Borehole D the dip of the strata indicates that this borehole is in the same limb of a major fold as the strata of borehole C. However, since the two sequences of strata show no direct correlation it must be assumed that a fault occurs running between boreholes C and D. (Naturally, there is insufficient evidence from just two boreholes to show which way the fault throws. However, we can give a minimum figure for the amount of throw.)

8

IGNEOUS FEATURES

Igneous features on geological maps may be divided into extrusive, sub-aerial and intrusive types. Extrusive features are lava flows, sub-aerial features are deposits of pyroclastic material and intrusive features are subterranean injections of magma into bodies of existing (country) rock. Although there are currently no active volcanoes in Britain, there are igneous features from the Precambrian to the Tertiary age, found mainly in the north and west of the country.

Lavas

These flow from volcanoes onto the surface. They will bury existing rocks and may be subsequently buried by later flows or sediments. As a result they will become interbedded with other strata, often conformably and thus concordantly. As such, lava flows will resemble sills (see intrusive features) and are difficult to distinguish from map evidence alone. Petrology may not help because small-scale intrusions and lavas may be composed of the same rock.

In the field it is possible to see a weathered upper surface to lava flows and there is contact metamorphism of rocks beneath the flow only: sills lack the weathering and bake the rocks above as well as below. There are other distinguishing features outside the scope of this book.

Lava flows are frequently composed of basalt and andesite because their lower viscosities enable them to flow over a considerable area.

Pyroclastic deposits

Pyroclastic material which falls into water will be mixed with detrital sediment, giving rise to a volcaniclastic sediment. More intense volcanic activity will produce beds of almost entirely pyroclastic material. Pyroclastic beds can also be deposited on land. Beds of consolidated material, called tuffs and agglomerates, may be conformably interbedded with lavas and sediments. For mapping purposes, most pyroclastic deposits can be treated in the same way as sedimentary strata.

Concordant intrusions

A majority of these are sills which are usually composed of basalt, dolerite or felsite. Sills are sheet-like bodies of magma, ranging from millimetres to many metres in thickness, injected between pre-existing beds of rock. They run parallel to the bedding direction and are said to be concordant. The intrusion seldom causes any observable disturbance of the strata so that a sill behaves structurally as though it were part of the stratigraphic succession. If there is subsequent tilting and folding of the strata the sill is also tilted and folded along with everything else. Sills in steeply dipping strata could be confused with dykes.

Using a map, a sill can be distinguished from surrounding strata in a few respects. It will, of course, have a different petrology to the surrounding rocks. It has also been intruded some time after the surrounding strata have been laid down. The strata may have already been displaced by faulting but the later sill may pass across faults without disturbance. Of course, faulting that post-dates both the strata and the sill will displace both.

A sill is generally concordant with surrounding beds but can change its horizon, appearing between different beds in different places. Often, such changes in stratigraphic position occur in abrupt steps because the magma has passed from one level to another using a joint or fault (Fig. 41). This can be observed in the Great Whin Sill of northern England.

The Whin Sill also shows another sill feature: changes in thickness. It varies in thickness from 30 m to 39 m in various places. A sill may also split into several 'leaves' and this can be seen in the Tertiary sills on the east side of the Trotternish Peninsula of the Isle of Skye. This phenomenon is not typical of lava flows although they may have smaller scale rafts of sedimentary rock caught up within them.

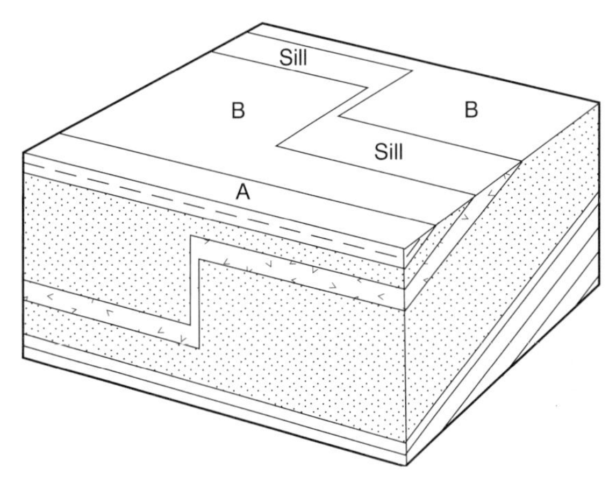

FIGURE 41 Block diagram showing a sill intruded into dipping strata. Note that the sill is seen to change horizon abruptly.

Composite sills (and composite dykes) occur when there is more than one injection of magma. The first usually forms the outer margins of the sill and a second, later, intrusion often runs through the centre, sandwiched by the earlier, solidified, magma. It is possible to see this on larger scale geological maps.

Sills are composed of resistant rock so they often cap hills, such as those of the Fife–Kinross border, protecting the softer rocks beneath. A sill may also form a pronounced escarpment, leading to waterfalls and providing a natural defence feature. Hadrian's Wall is partly constructed along the Whin Sill and Stirling Castle sits upon a sill escarpment.

Other concordant intrusions, such as laccoliths and lopoliths, are rare and not covered here.

DISCORDANT INTRUSIONS

Dykes

These are vertical or steeply dipping sheet-like intrusions, from a few millimetres to several metres thick (rare examples reach tens of metres across). They can often be traced across country for many kilometres.

Since they are approximately vertical, dykes follow straight paths across the landscape regardless of changes in topography. It is sometimes possible to date a dyke relative to the surrounding sedimentary rocks. It must be later than the youngest beds through which it cuts and, if it lies beneath an unconformity, will be older than the beds above the unconformity if it does not cut through these beds (cf. the dating of faults, p. 47).

Dykes are intruded during periods of crustal tension. Widening cracks or joints permit magma to be intruded both from below and laterally. Dykes may be very numerous near to volcanic centres, where they may indicate a considerable extension of the crust. The dyke swarm on the south coast of the Isle of Arran is a good example (Map 29).

Ring-dykes

These are circular or arcuate dykes surrounding a volcanic centre and usually composed of higher silica magmas. They are near vertical in section and are associated with down faulting when a volcanic centre undergoes late-stage subsidence. Ring dykes can reach tens of metres in thickness.

Cone-sheets

These are also circular in plan and form discontinuous arcs but dip steeply inwards towards the volcanic centre. They occur in swarms and a notable example can be seen at the Ardnamurchan peninsula in Scotland. Cone sheets are generally thinner than ring dykes, being only a metre or two in thickness. They are associated with earlier stages of volcanic activity and are frequently basaltic in composition.

Stocks, bosses and batholiths

These are larger plutonic intrusions, often of granite, emplaced deeper within the crust. They become exposed by prolonged erosion. Near-circular stocks and irregularly shaped bosses are usually tens of kilometres across. Batholiths are larger intrusions which can reach hundreds of kilometres in size. The Dartmoor granite is one of several stocks protruding upwards from a batholith that extends under much of the South West Peninsula.

These discordant intrusions are often longer than they are broad because they follow the structural geology of the region, but in every case they have steep margins which cut through the surrounding and pre-existing strata. Because they are composed of relatively resistant rock, large intrusions can form upland areas. They are surrounded by a zone of rocks which have been altered by heat from the magma. These contact metamorphic aureoles are typically a few kilometres across at most. Smaller intrusions, such as sills and dykes, affect only a few centimetres or metres of the rock surrounding them. Large intrusions also cause extensive mineralization. Mineral veins are often dyke-like on maps because they fill vertical fissures.

Volcanic necks

These result from the infilling of volcanic vents with consolidated lava (basalt, etc.) or pyroclastic material (vent agglomerate, tuff, etc.). They cut through existing strata, have near-vertical sides, and they are usually circular in plan. A volcanic neck has the same shape as a boss but is much smaller and is composed of different rocks.

Using a map, it is sometimes difficult to distinguish between a volcanic neck and a near-circular hill-top cap of igneous rock (a sill or lava flow). In section, however, they look very different (Fig. 42).

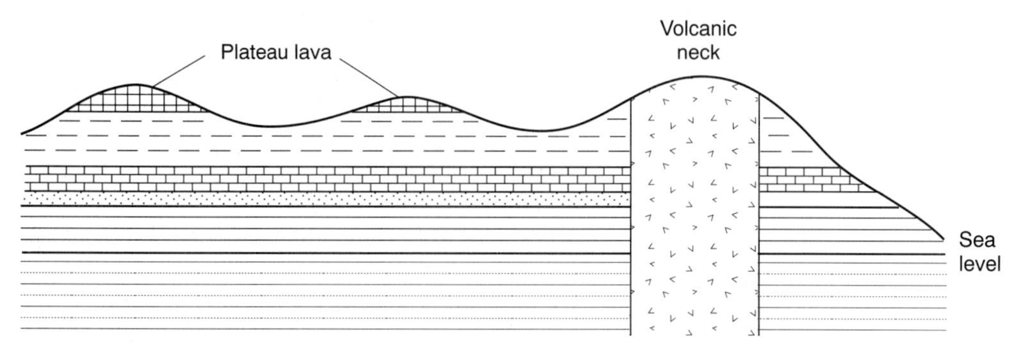

FIGURE 42 Geological section showing a volcanic neck and lava-capped hills.

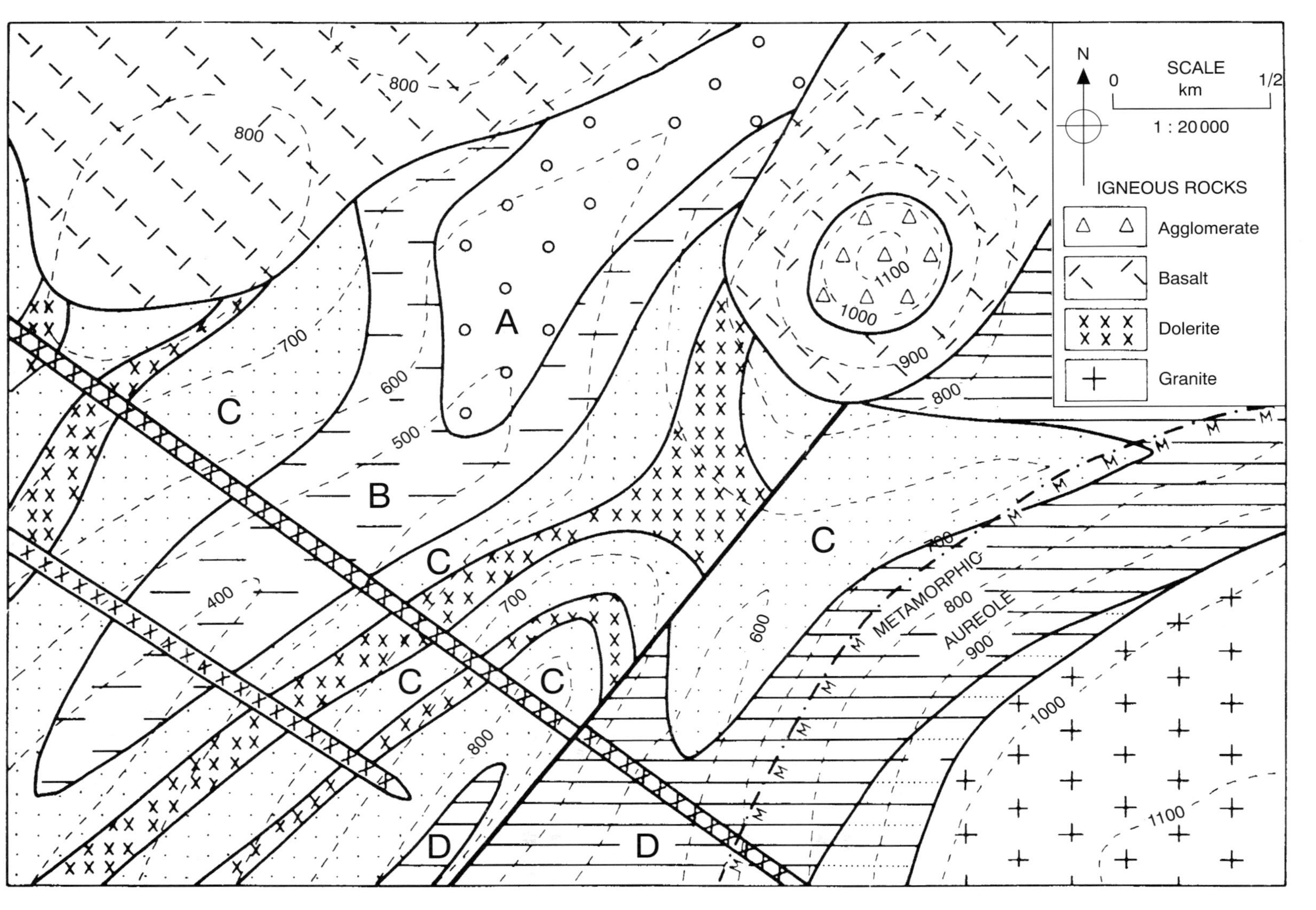

Map 26 Draw sections across the map to show the form of the igneous rocks and the other structural features. Also deduce the relative ages of the igneous rocks as far as this is possible.

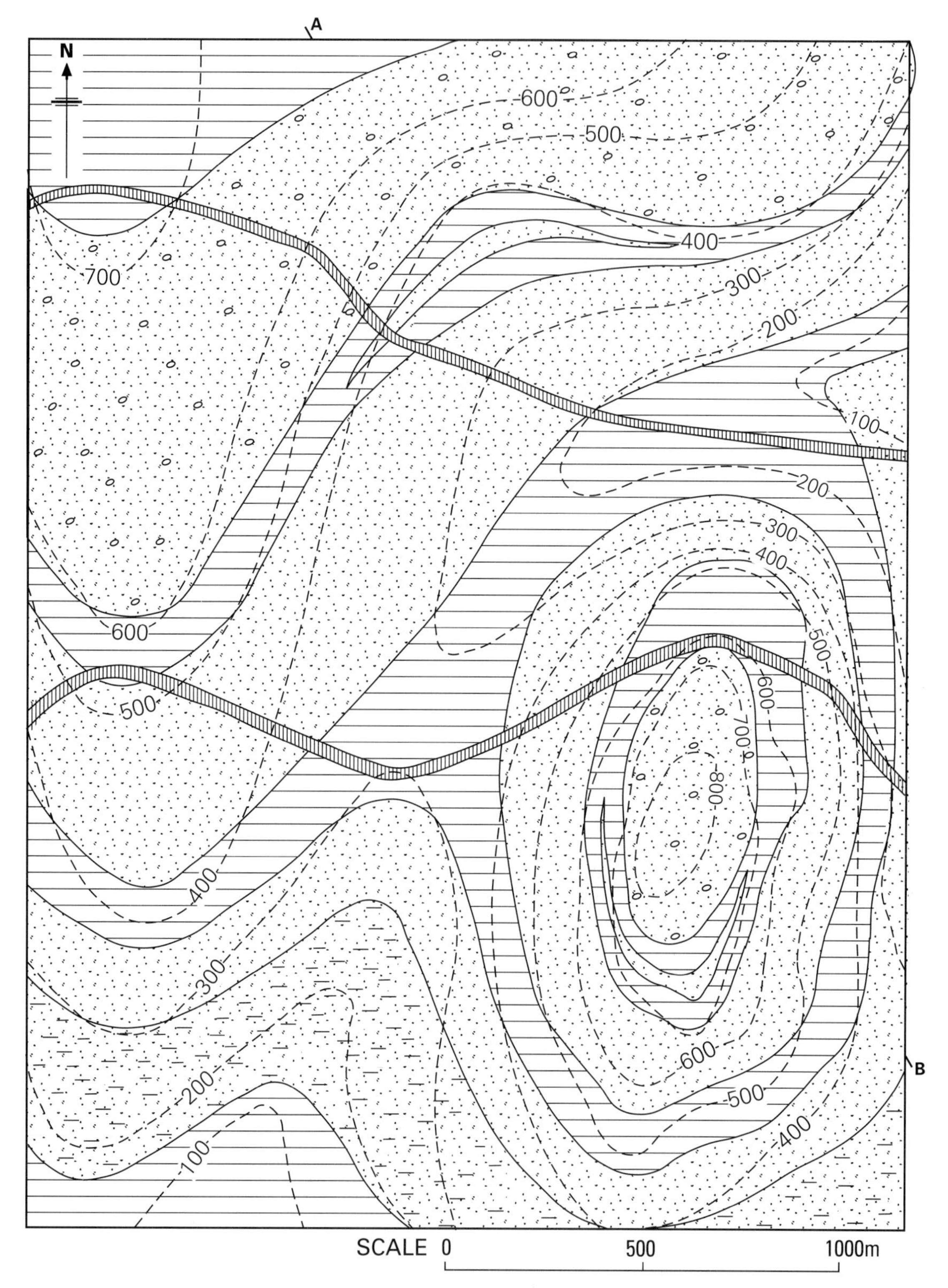

Map 27 How many dykes and sills are visible on this map? What criterion have you used to distinguish the sills from the dykes? What evidence is there against the sills being lava flows? Draw a section from A to B using the profile provided. Calculate and comment on the dip of the dykes.

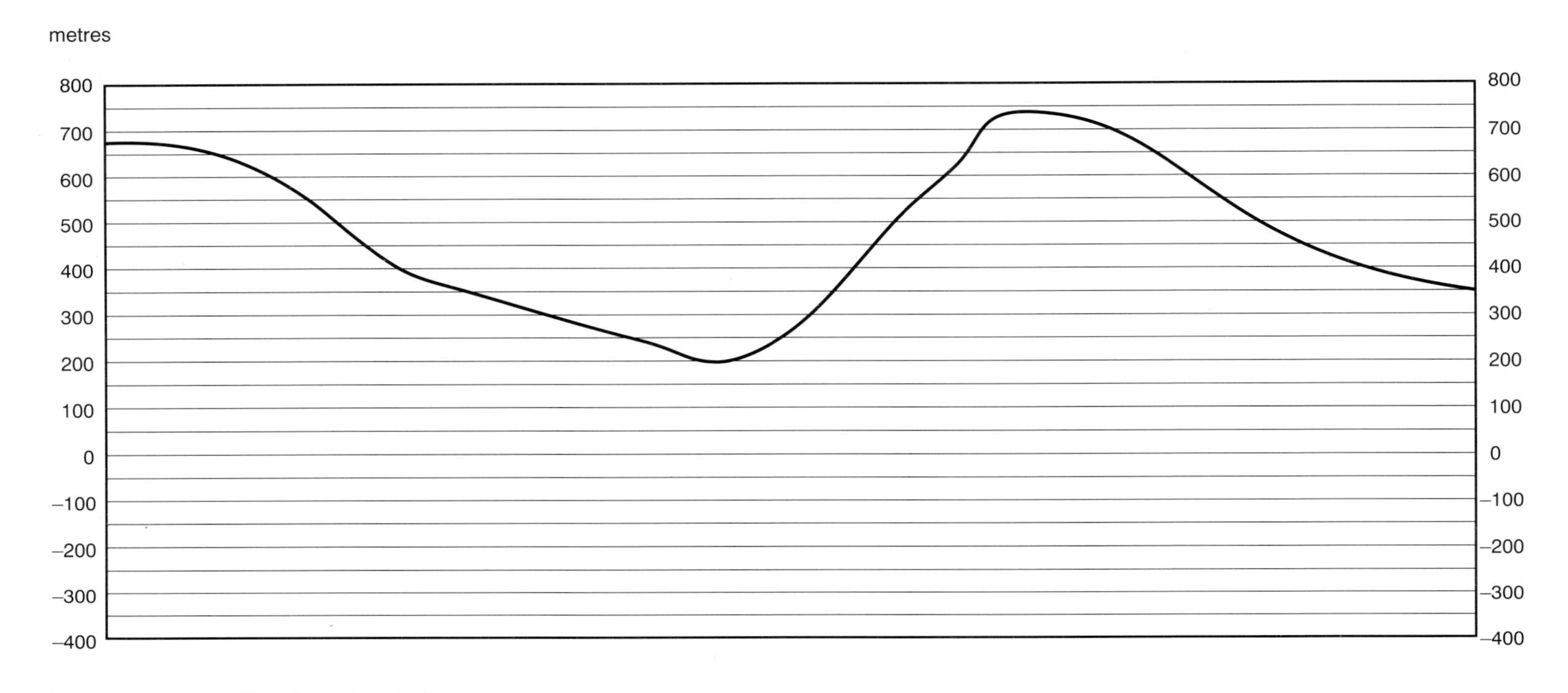

Topographic profile of section A–B.

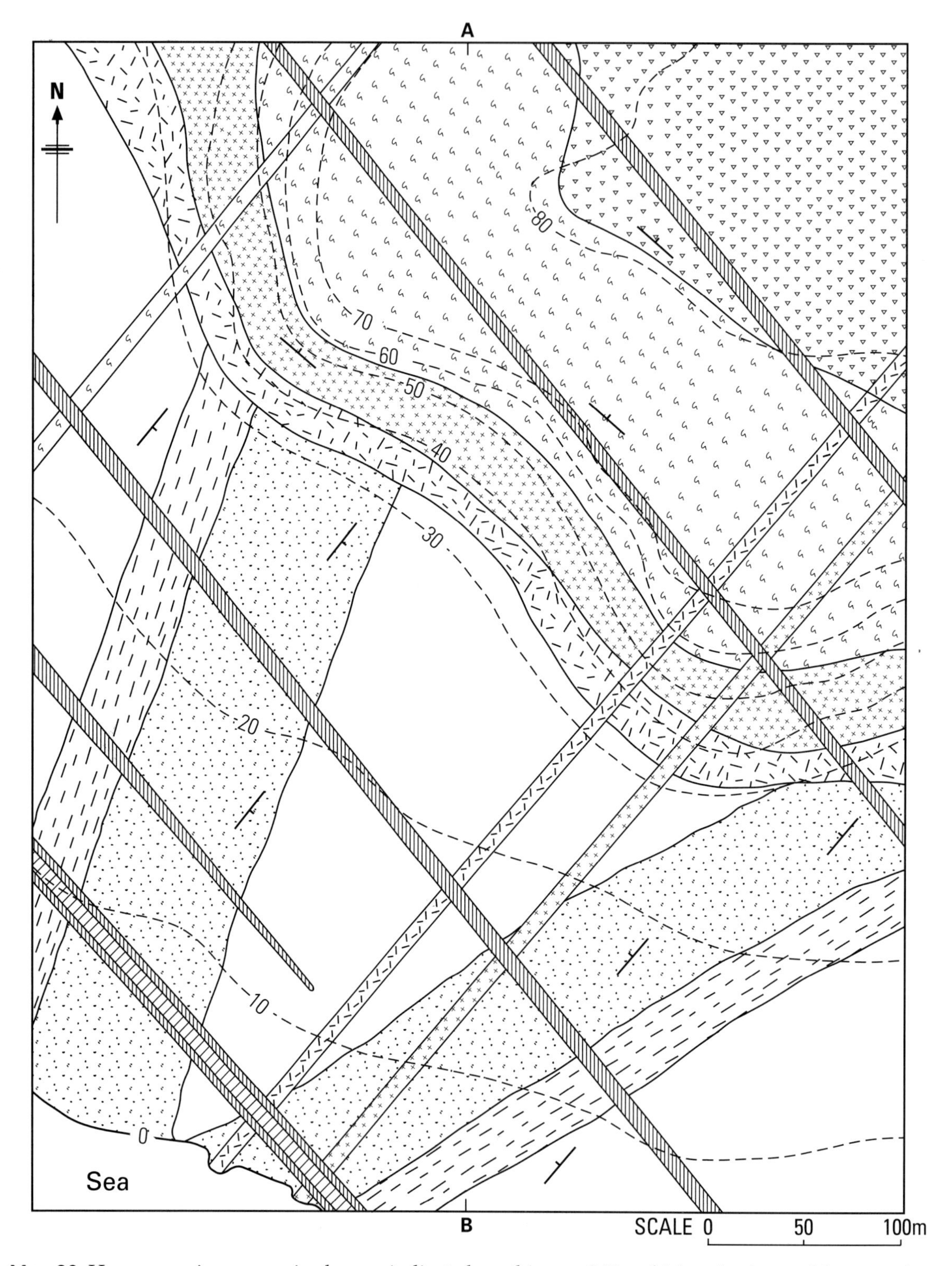

Map 28 How many igneous episodes are indicated on this map? To which episode would you assign the composite dyke? Indicate an unconformable boundary on the map (other than dyke boundaries). Draw a section from A to B. Calculate the dip of the lavas and tuffs. What is the probable dip of the dykes?

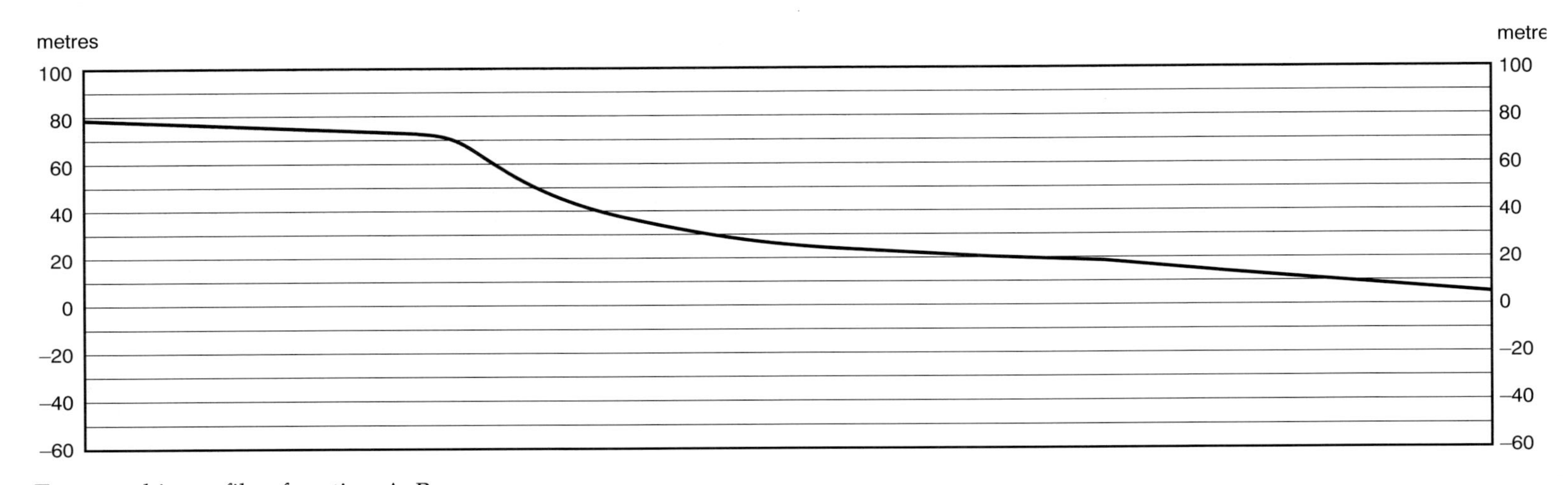

Topographic profile of section A–B.

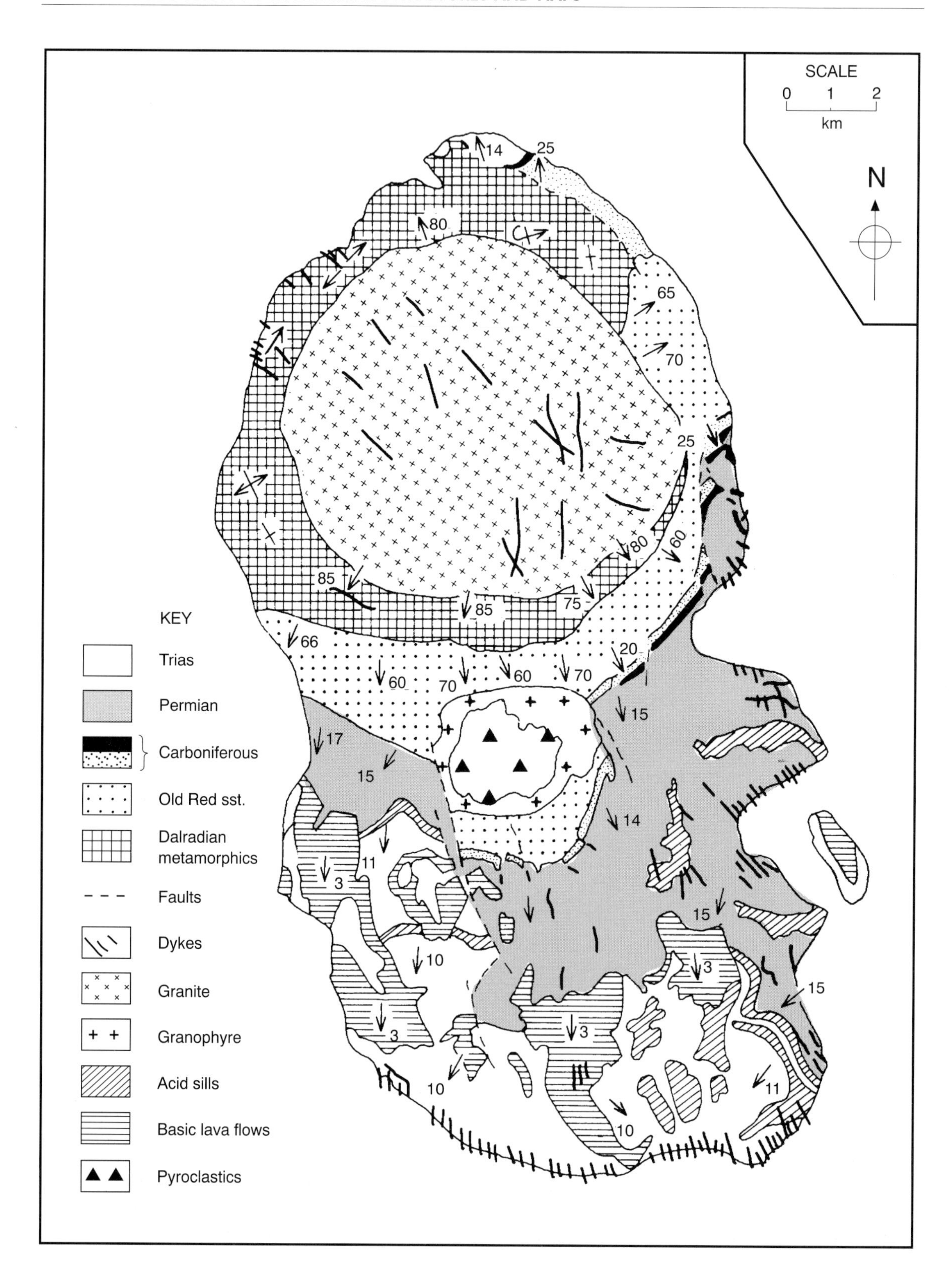
SCALE
0 1 2
km
N
KEY
Trias
Permian
Carboniferous
Old Red sst.
Dalradian metamorphics
Faults
Dykes
Granite
Granophyre
Acid sills
Basic lava flows
Pyroclastics

Exercise on Geological Survey Maps

1. **Edinburgh: 1″ Map No. 32 (Scotland).** Give the relative ages of the various types of igneous intrusion shown on the map. How can the relative age of an igneous rock be discovered from map evidence and what are the limitations of this method? What other evidence might one expect to find in the field but which cannot be discovered from a map? (This map has now been replaced by 32E Edinburgh and 32W Livingstone on the 1:50,000 scale.)

Map 29 This map is broadly adapted from the geological map of the Isle of Arran 1:50,000 Special Sheet. Most of the kinds of igneous intrusion shown on the stylized Map 26 can be found here. Write an account of the geology of the area, noting particularly the two major unconformities, the dips of strata of different ages and the distribution and disposition of different igneous rocks. What can you deduce from the general direction of the dykes? It is possible to work out the geological history for this map (however, it will not be identical to that of the one-inch to the mile geological survey sheet).

9

PLANETARY GEOLOGY

With the growth of space probe imaging, geological mapping principles have become useful for interpreting the surfaces of the rocky inner planets and the icy moons of the outer planets. This chapter provides simplified examples based on real surface images. It should be remembered that the solutions cannot be supported by detailed surface exploration. There are always competing theories to explain some of the features seen.

The principle of cross-cutting features described in the preceding chapter on igneous features can be applied to planetary surfaces: new features will cut across old ones anywhere. Unlike the Earth, many solar-system bodies have largely unmodified surfaces showing ancient cratering. During early solar system times there were numerous impacts of fast moving comets and meteorites onto the surfaces of the newly forming planets. With time the number of objects capable of potential collision has decreased. On any particular body, the greater the crater density the older the surface is likely to be. Craters are obliterated by reworking to produce younger surfaces. This is usually an internally driven process but planets with atmospheres may also have experienced surface erosion and deposition by wind or water.

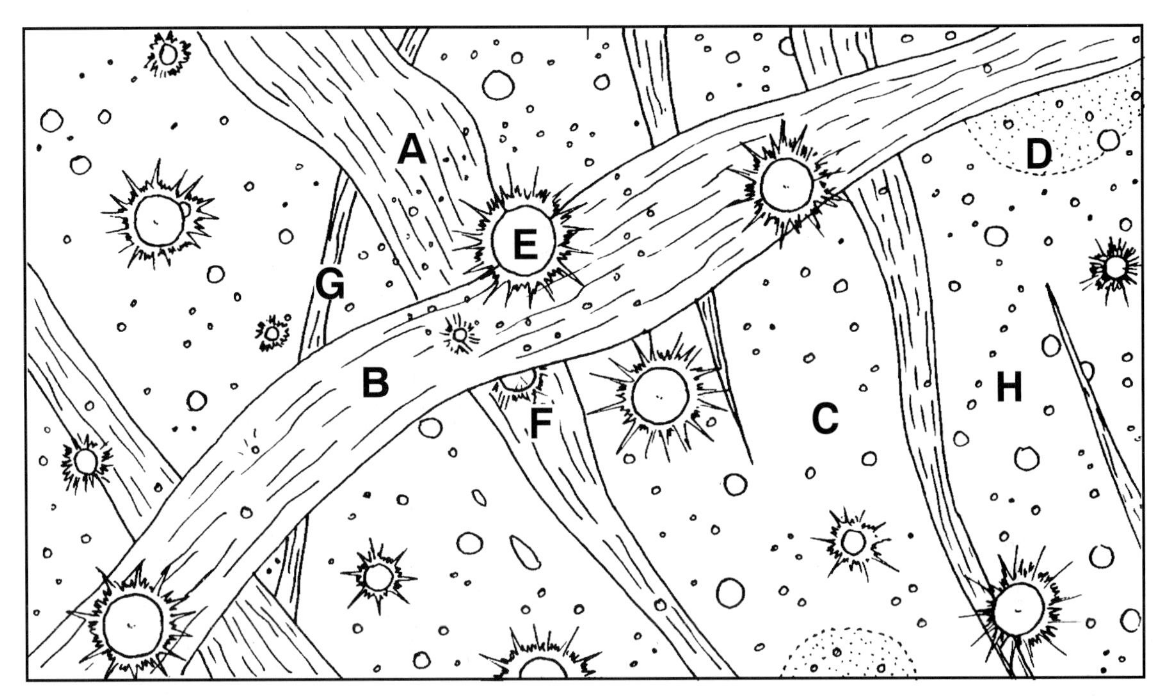

MAP 30 The surface of Ganymede is represented by a line drawing. The real surface is composed of ice which has become darkened by contaminants with time. The linear features are paler and the youngest craters and their ejecta appear white. There is virtually no atmosphere so surface processes are minimal.

1. There are craters of various sizes. Suggest factors that will influence their dimensions.
2. Some of the craters are surrounded by ejecta blankets. What factors will influence the extent of these blankets? Why is the ejecta material white and why are these craters paler than the others?
3. Why is crater E surrounded by a 'halo' of smaller craters?
4. Suggest a cause for the pattern of craters at H.
5. The circular feature D has largely disappeared. What might it have been and what may have happened to it since? Refer to the next question for a clue.
6. The large linear features are cracks between crustal blocks. Suggest what has happened to the crust and a mechanism for it.
7. Of its four larger moons, Ganymede is the third most distant from Jupiter. The second moon, Europa, has a cracked icy surface with very few craters, whereas Callisto has no cracking but extremely dense cratering. Suggest how their surfaces have evolved compared to Ganymede and place the surfaces of these three moons in age order.
8. Arrange the surface features A to F in age order using the principle of cross cutting.

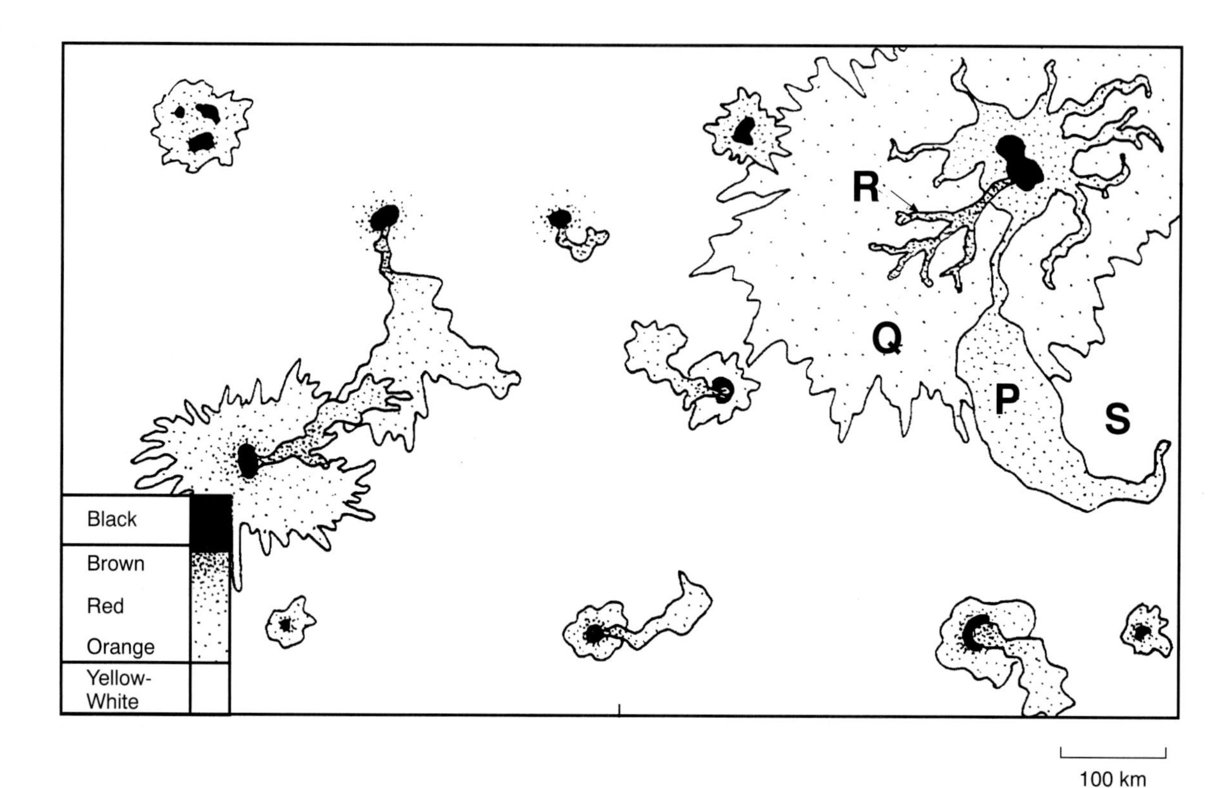

MAP 31 The surface of Io. This is the innermost of Jupiter's major moons and has a silicate crust containing sufficient sulphur and sulphur dioxide to colour the surface. Sulphur and sulphur dioxide are yellow/white when cold and solid, become orange on melting and progressively darken as they approach boiling point. Due to the gravitational forces of Jupiter and the other major Jovian moons, Io has considerable internal energy to drive geological processes.

1. Based on sulphur colours, describe which parts of the surface map are hottest and which are coldest.
2. Identify the brown/red curving features and explain why they progressively change colour.
3. Identify the black features.
4. Suggest an age order for features P to S.
5. There are no impact craters on Io's surface. Suggest why this should be.
6. Using the answer to question 7 about Ganymede, place all four major Jovian moons in age order. How does this relate to their distance from Jupiter?

MAP 32 The map is based on the Chryse Planitia region of Mars. The surface is composed of rock beneath a thin atmosphere of carbon dioxide. Surface processes are important here.

1. Compare the density of impact cratering to the surface of Ganymede in the earlier map. Suggest reasons for a difference.
2. The craters have tails of fine sediment. Comment on their direction and how they might have formed. Are these features likely to be contemporaneous with the cratering or more recent?
3. Linear features in the sediment are visible on the upper left of the map. Comment on their likely origin and their relationship to the crater sediment tails. What has happened to the craters adjacent to the linear features?
4. A wide shallow channel runs from top to bottom of the map. Within it there are raised teardrop features. Comment on the most likely agent of erosion that formed this channel.
5. Which way did the channel probably flow? Identify the features that led you to this conclusion.
6. The channel is ancient and none are currently forming on Mars today. Suggest reasons why not.
7. Comment on the most likely agent of erosion that caused the sinous features at the bottom left of the map. Where does the eroding agent appear to come from.
8. Arrange the major channel, sediment features and craters in correct time order.

SOLUTIONS TO PLANETARY GEOLOGY MAPS

Ganymede

1. There are several factors that will influence crater dimensions. Since kinetic energy is proportional to the mass and the square of the velocity, the mass of a meteorite and its speed relative to the impact surface are both important. The strength of the surface material may also have an effect: ice at low temperatures behaves in a similar way to rock. The angle of descent of the impacting object will alter the crater shape (see also answer 4).
2. Ejecta blankets consist of material thrown out of the crater during the impact. Apart from the factors already mentioned in answer 1, there is gravity to consider. Ejecta will travel further in a lower gravity environment. The craters gouge through the darkened surface material to reveal fresher ice beneath, some of which is thrown out: hence the colour of the ejecta. Younger craters reveal fresher material which has not darkened yet.
3. Crater E is surrounded by smaller craters where small ejecta objects have come to rest on the surface.
4. This is a chain of craters caused by an impacting object that has repeatedly bounced along the surface before stopping.
5. This is an ancient (large) crater which has gradually flattened out due to gradual ice flow (see also next answer). Such craters are appropriately termed palimpsests.
6. The crust has, at some time, broken into blocks which appear to have moved on a liquid layer beneath before refreezing with slushy looking fissures between the blocks. This is analogous to pack ice. The energy source may have been radionuclides in the rocky core of Ganymede.
7. The extensive cratering of Callisto suggests that its surface is older (over 4000 Ma) compared to Ganymede, whereas the almost craterless surface of Europa is thought to be much younger (about 100 Ma).
8. Based on cross cutting, the correct sequence is C (the original and more extensively cratered surface), D, H, G, A, F, B and E.

Io

1. The hottest regions are localized and likely to be volcanic vents. The areas surrounding them are progressively cooler and the 'plains' between the volcanoes are coolest of all.
2. The curving features look like 'lava' flows containing at least some liquid sulphur. They change colour as they cool away from their source, the vents.
3. Volcanic craters. The black material is not only very hot liquid sulphur but also erupting plume material.
4. Progressive sulphurous flows overlie each other, hence the age order is S, Q, P and R.
5. The lack of impact craters evidently reflects the extreme activity which is constantly resurfacing the crust.
6. The younging sequence is Callisto, Ganymede, Europa and Io. This is also the order of decreasing distance from Jupiter. What is the connection?

Mars

1. There are fewer impact craters for several possible reasons. Mars is in a different part of the Solar System where there may have been fewer impacting objects. Smaller projectiles will be 'burned up' upon entry into the Martian atmosphere. The craters may have been destroyed by internal processes or eroded or covered up by external processes. External processes are evident on this map.
2. The fine sediment tails are probably wind blown dust accumulating in the lee of each

crater. Their NNE direction suggests a wind blowing from the SSW. Sandstorms are a regular feature of current Martian weather so we can assume that the tails are fairly recent, unlike the craters.

3. The linear features are probably dunes of wind-blown sediment. The orientation of the dunes suggests that they are formed by the same winds as the sediment tails, although, without the additional evidence from the tails, it is not certain whether they are advancing NNE or SSW. Several craters are wholly or partially buried by the dunes but it is reasonable to assume that those on the southern margin are becoming slowly uncovered.
4. This is probably a large braided river channel, although it is now dry. Water is likely to have been the agent of erosion.
5. The teardrop features often form around craters where the ground has been fused by the impacts. The tails of such features point downstream in Earthbound situations so it is reasonable to assume that the river once flowed from the bottom to the top of the map.
6. The general consensus is that the atmosphere was once sufficient to sustain liquid water on the surface of Mars. This planet has probably lost most of its early atmosphere and has also dried out. Water ice can still be found at the poles and may exist as permafrost elsewhere.
7. This is another water-made feature. Catastrophic floods may have been caused by the melting of permafrost.
8. In time order we can place the cratering as the earliest feature, followed by the water erosion features (such as the braided channel) and finally the wind blown features (the sediment tails and linear dunes).

10

ECONOMIC PROBLEMS

In a book devoted to map problems and their solution, many aspects of economic geology must be left to other textbooks. However, the reader should appreciate that problems in economic geology are dependent upon the analysis of geological structures and in many cases economic calculations are made on data derived from geological maps. We have seen the simplest application of an economic problem already in Map 10, the deduction of the position of the sub-unconformity 'outcrop' of a coal-seam from structure contour intersections. Simple economic problems were posed on Maps 16 and 17.

Here are seven maps, ranging from simple to advanced, dealing with coal seams, ironstones and an ore-body. The problems include drawing structure contours, overburden isopachytes and bed-thickness (stratal) isopachytes.

MAP 33 The map shows an area of considerable relief with contours at 100 m intervals shown by broken lines. A thin but persistent oil shale outcrops and is cut by a dip fault with a downthrow of 200 m to the east. The structure contours for the oil shale, deduced from dip measurements on the outcrop and confirmed by borehole information, are shown by fine lines. Draw the isopachytes for the cover (= overburden) overlying the oil shale at 100 m intervals (100 m, 200 m, 300 m, etc.). Remember that an isopachyte joins up points of equal thickness. The difference between the height of the ground and the height (or depth) of the oil shale at any intersection of the ground contours and the structure contours will give the thickness of the overburden at that point. All figures on the map are given as heights in metres relative to sea level.

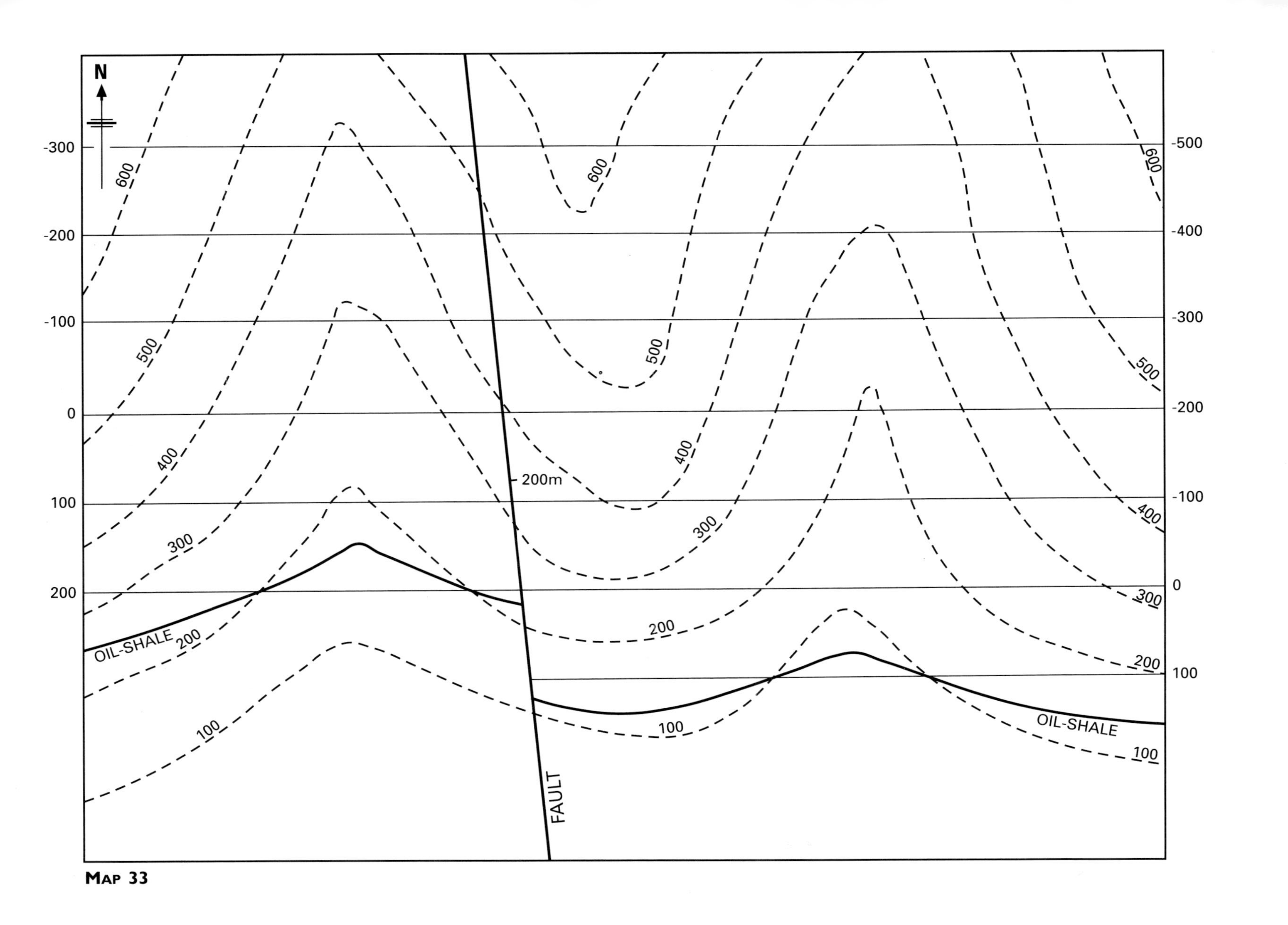
N
-300
-200
-100
0
100
200
-500
-400
-300
-200
-100
0
100
600
500
400
300
200
100
200m
OIL-SHALE
FAULT

Map 33

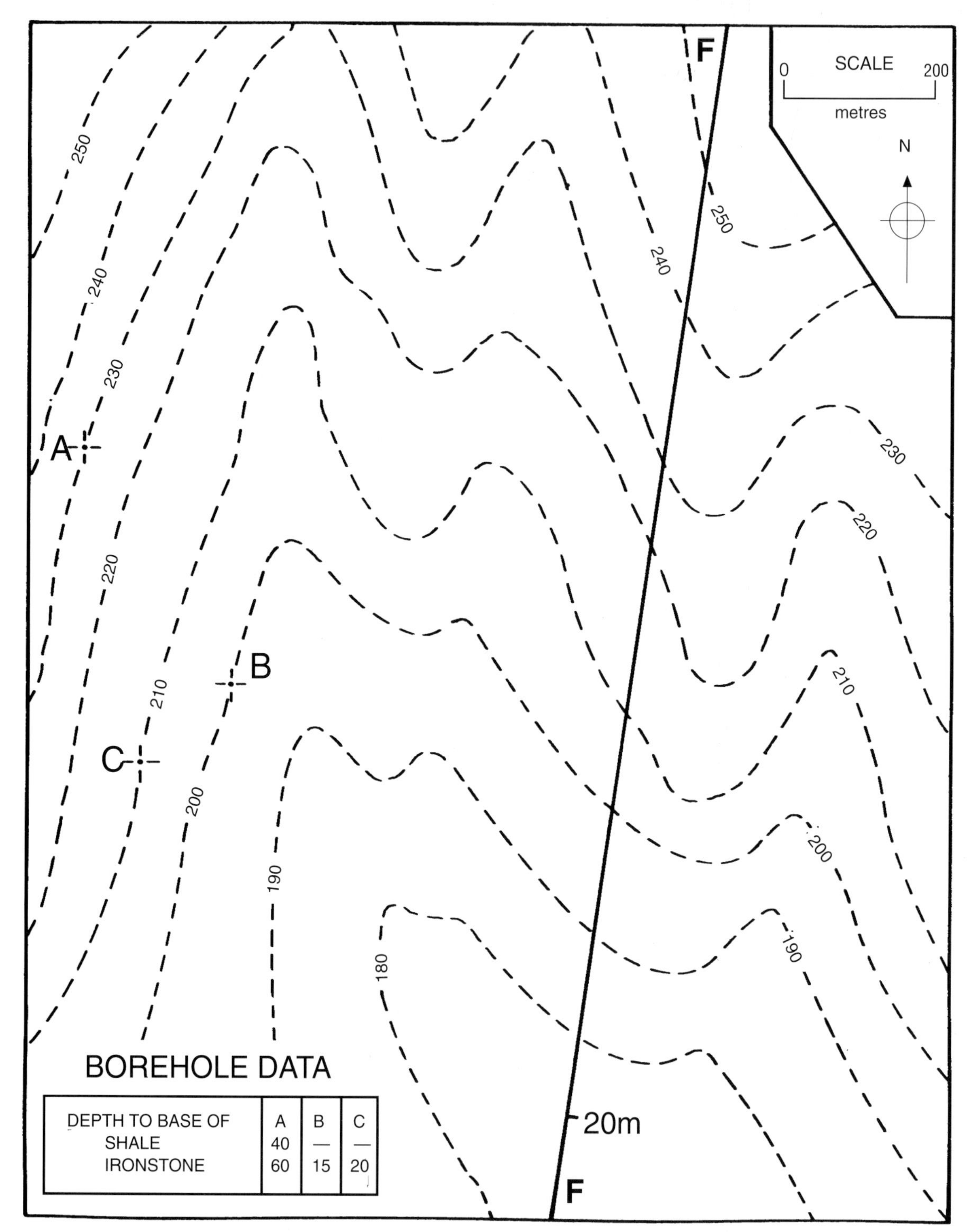

Map 34 The western part of the map comprises a three-point problem enabling us to draw structure contours at 10 m intervals on the base of the ironstone. Assuming the top of the ironstone to be 20 m higher, why does borehole B penetrate only 15 m of ironstone? East of the fault (a normal fault of high angle of dip), produce the structure contours and re-number them 20 m lower (since the fault is shown as having a downthrow of 20 m to the east). Shade outcrops of iron ore on both sides of the fault. Also shade areas where the ironstone could be worked opencast if not more than 40 m of overlying shale is to be removed, i.e. draw the overburden isopachyte for 40 m. Draw a section along the North–South line N–S passing through point B and another section along the east–west line E–W passing through B.

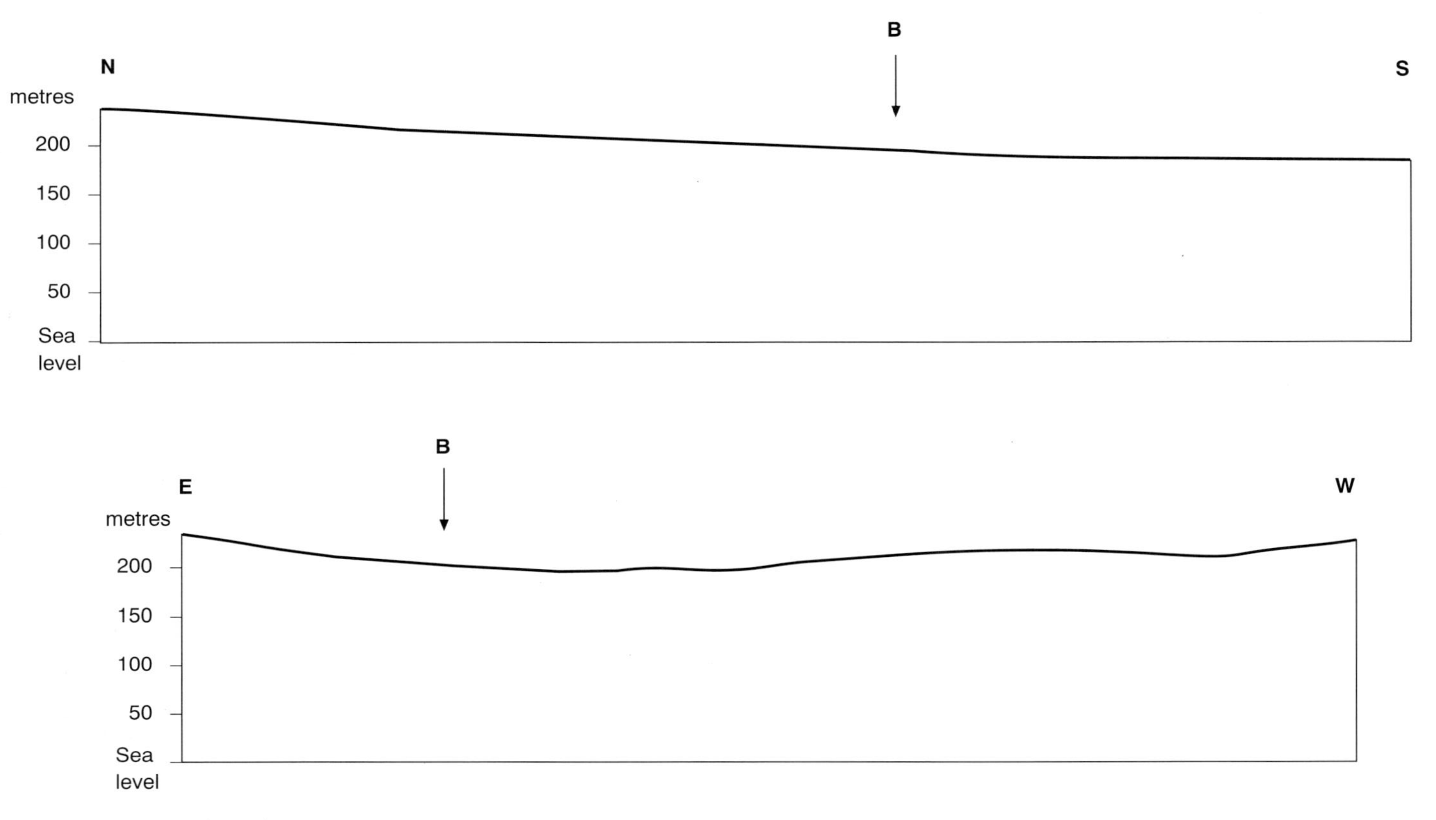

Topographic profile of sections N–S and E–W.

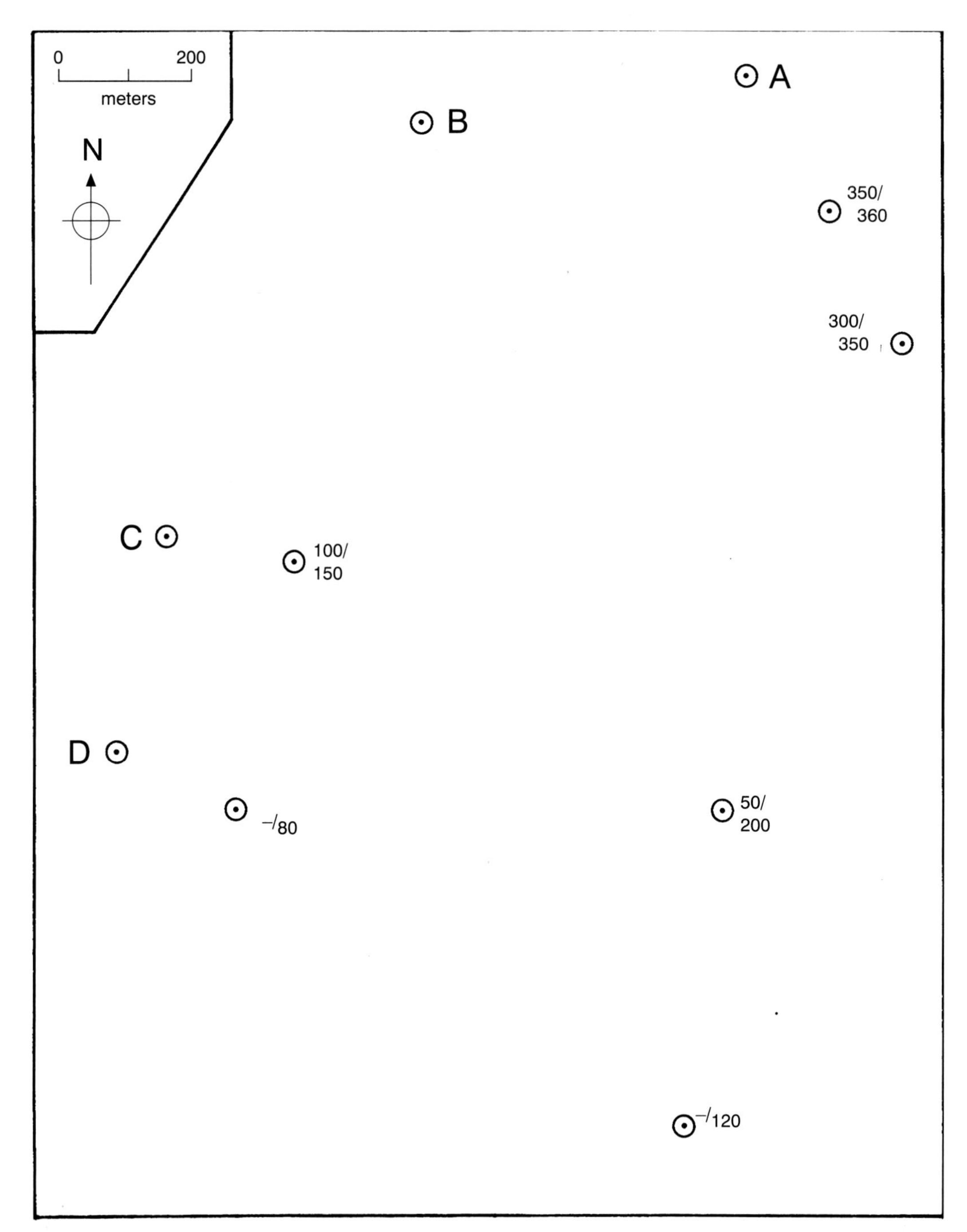

MAP 35

MAP 35 A wedge-shaped, igneous, sill-like body is encountered in boreholes. The land surface is flat and horizontal over this area and the depths from the surface to the top and base of the sill are given beside each borehole. Construct structure contours for the top of the sill and its base, assuming that both top and bottom are uniformly dipping (but not parallel). Construct isopachytes at 50 m intervals to show how the sill varies in thickness. Shade the area where the sill outcrops. Why is the sill absent in boreholes A and B? If it was not found in boreholes at C and D how would you explain its absence?

Note on Map 35. The reason for the absence of the sill in some boreholes could be as follows: (a) it outcrops so that at the site of the borehole it is no longer found as a result of erosion, (b) the ore-body thins to nothing and is not found, (c) if the absence of the sill is not due to its being eroded away nor to its thinning, it must be absent in a borehole due to faulting. (The ore-body may have been upfaulted then removed by erosion or it may have been downfaulted to such a depth that the borehole was not deep enough to find it. There is no map evidence given here to indicate which explanation is the more probable.)

MAP 36 The outcrops of a coal-seam have been revealed in an area of flat ground. It can be seen that the lateral displacement of the *outcrop* is a consequence of a normal fault with a 75 m throw. The seam was encountered in four boreholes, A to D, at depths given on the map showing that it has a uniform dip in a generally north-easterly direction (37°E of N). The coal thins from a maximum of over 6 m in the south-east to less than 2 m in the north-west. The isopachytes on the map show that this thinning is not uniform. It is proposed that it would be economical to mine the coal by opencast methods (= strip mine) to a depth of 150 m where the seam is not less than 3 m thick. Draw structure contours on the top surface of the coal-seam at intervals of 50 m and shade on the maps the areas where the coal may be economically mined.

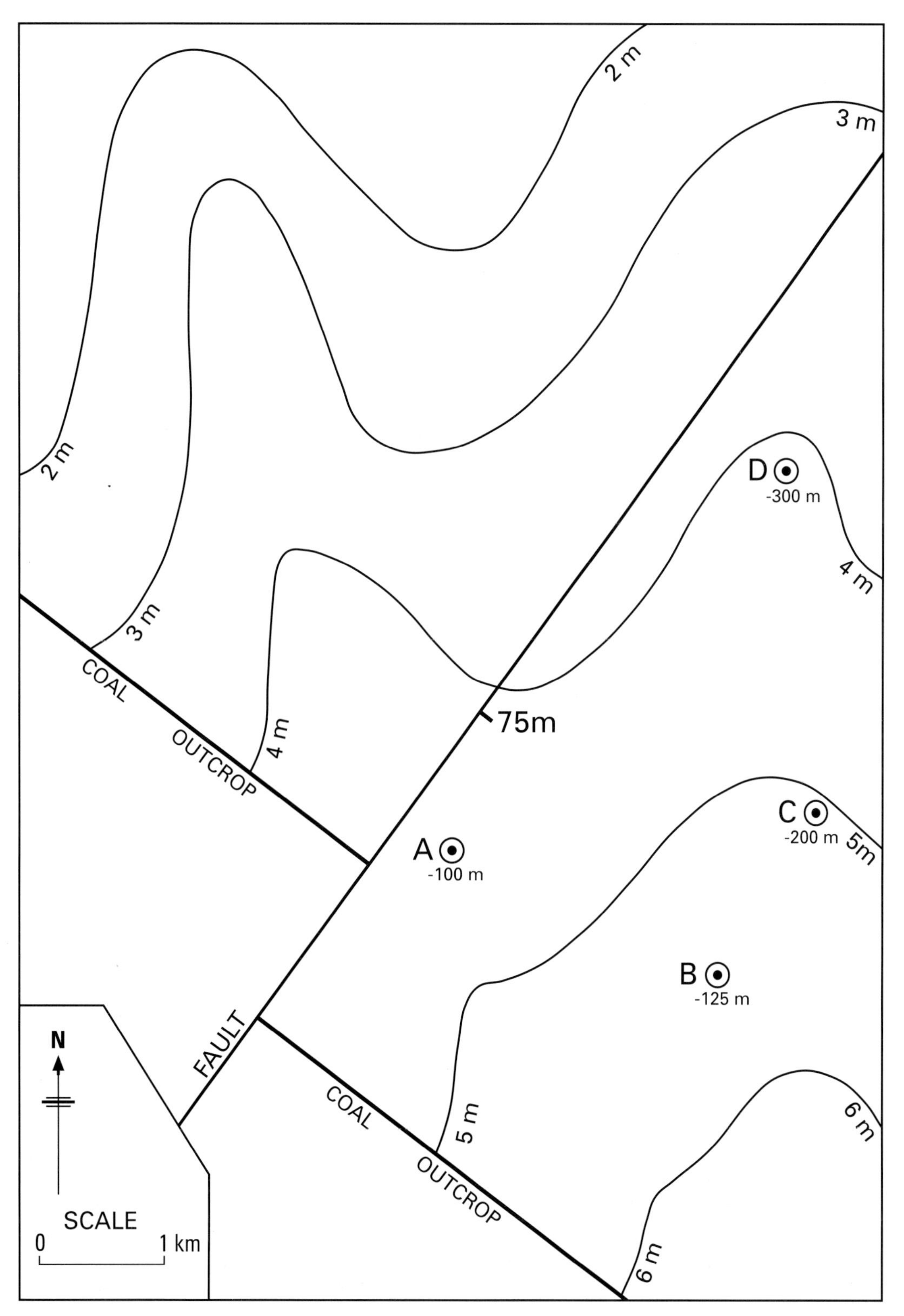
2 m
3 m
2 m
3 m
4 m
D
-300 m
4 m
COAL
OUTCROP
4 m
75m
C
-200 m
5m
A
-100 m
B
-125 m
FAULT
N
COAL
OUTCROP
5 m
6 m
6 m
SCALE
0
1 km

Map 36

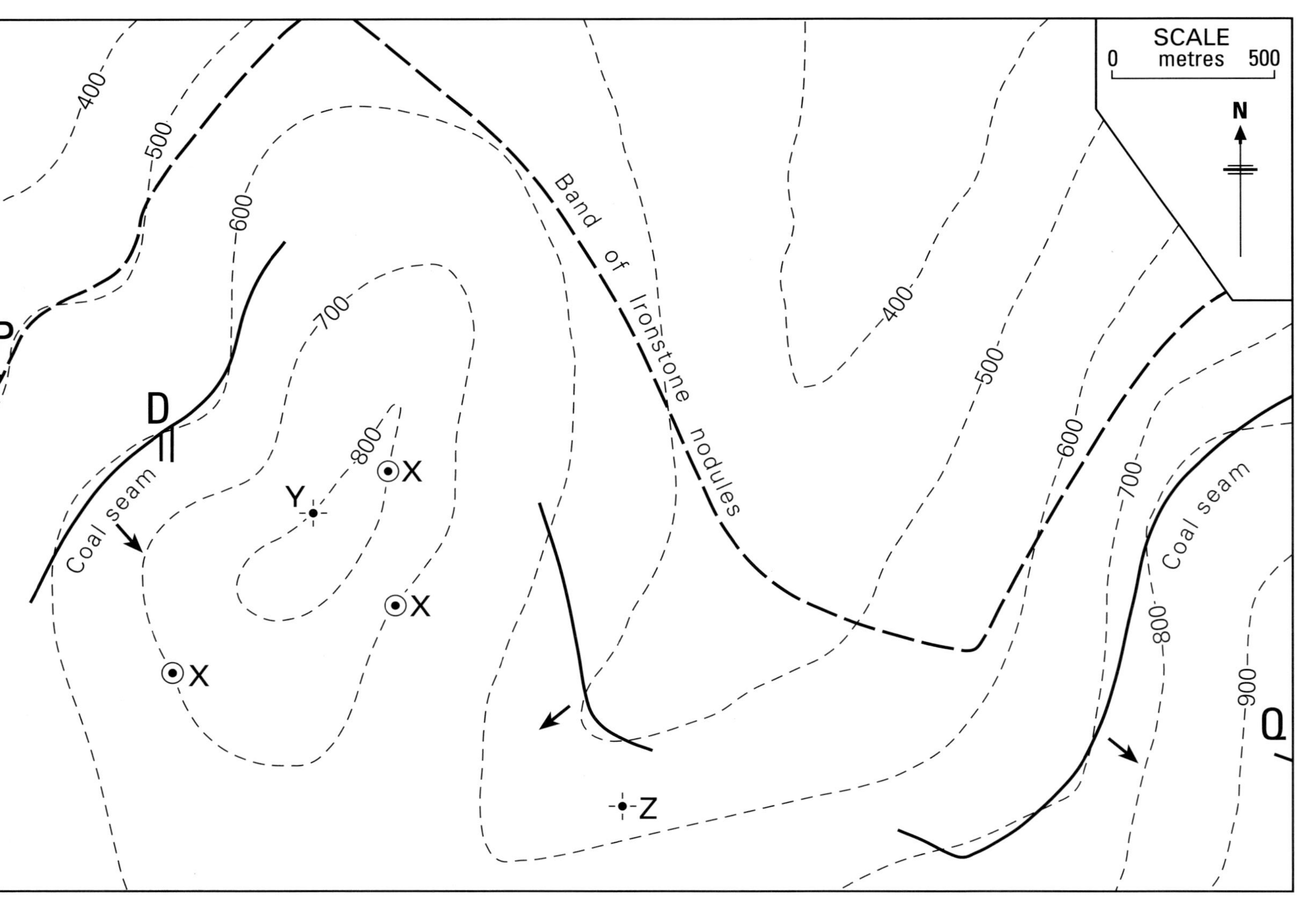
SCALE
0 metres 500
N
Band of Ironstone nodules
Coal seam
Coal seam
P
Q
D
X
X
X
Y
Z
400
500
600
700
800
900

Map 37

MAP 37 The map shows part of the outcrop of a coal-seam. It was also found at a depth of 300 m in the boreholes at several points marked X. Complete the outcrop of the coal-seam. Determine the depth at which it would be encountered in shafts put down at Y and Z. Also insert the outcrop of another seam which lies 300 m higher (vertically) in the succession. What is the amount of plunge of the fold axes? Insert outcrops of the fold axial planes on the map. Draw a section along the line P–Q. Suppose an adit (= drift mine) was situated at D. Naturally, this tunnel follows the slope of the coal-seam, although it is running due south. What would be the gradient of this tunnel? NOTE: What you are calculating is the apparent dip of the coal on bearing 180 (expressed as a gradient).

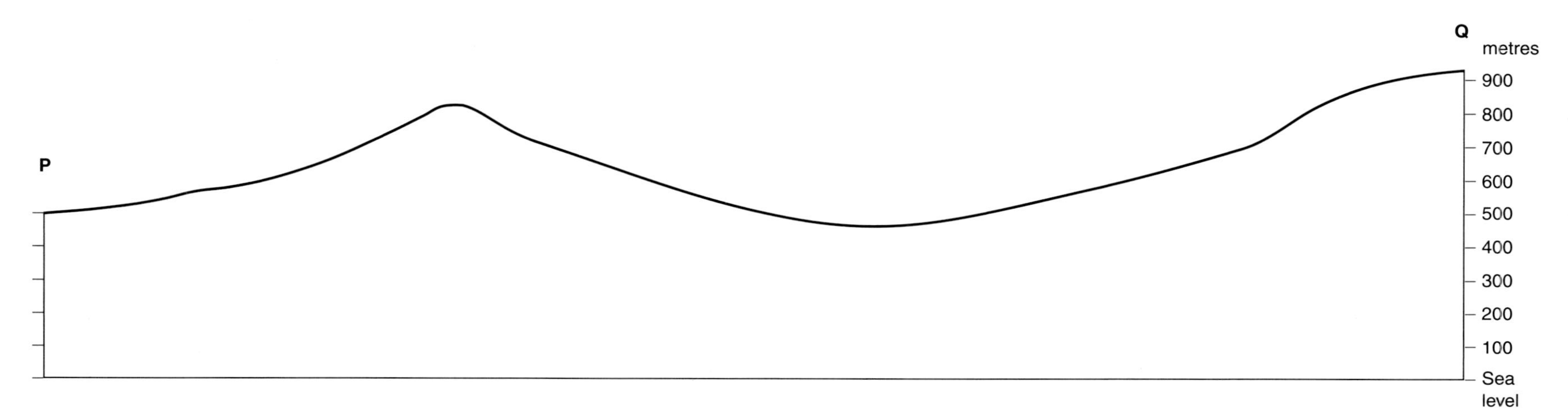

Topographic profile of section P–Q.

JOIN

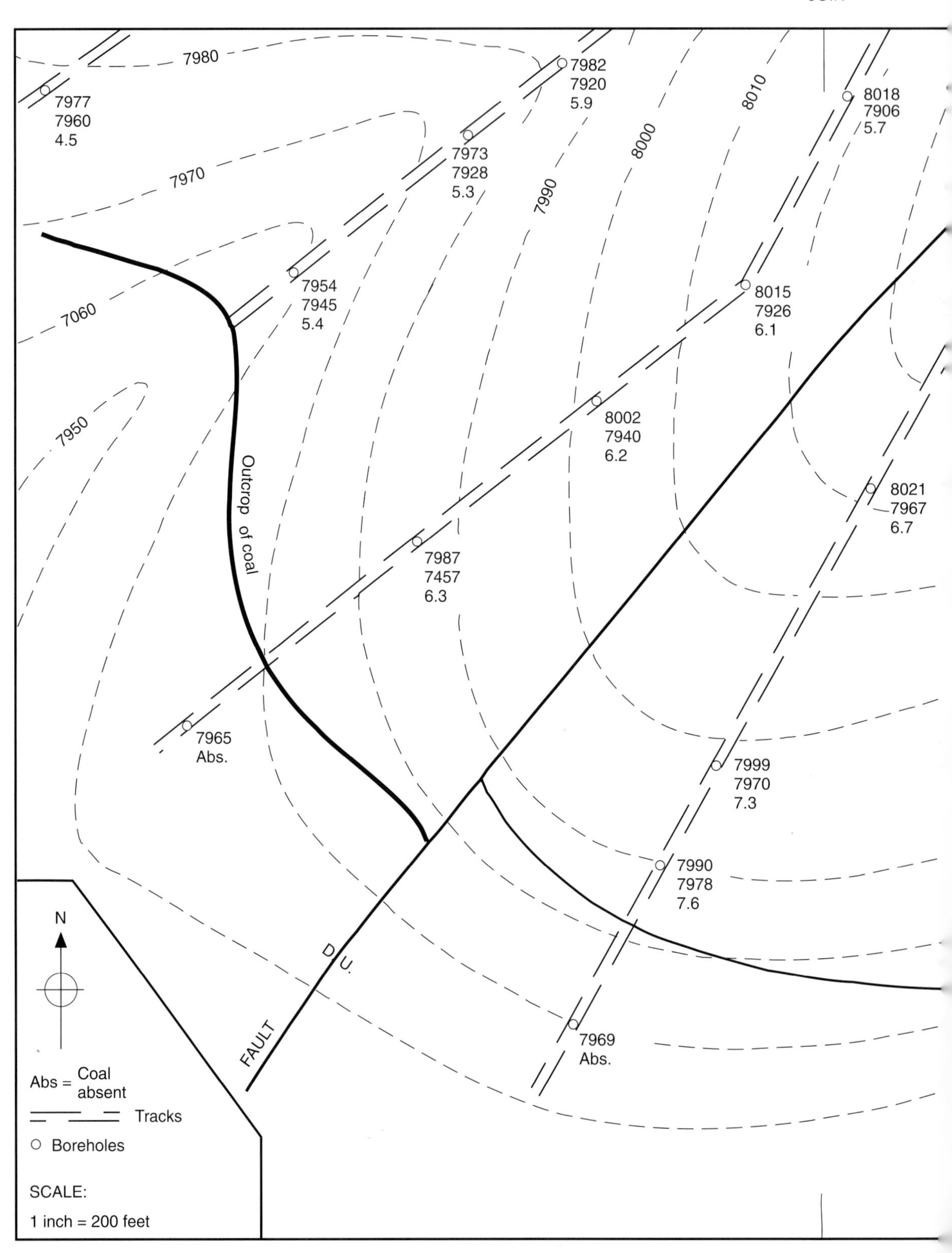
7980
7970
7060
7950
7990
8000
8010
7977
7960
4.5
7982
7920
5.9
7973
7928
5.3
7954
7945
5.4
8018
7906
5.7
8015
7926
6.1
8002
7940
6.2
8021
7967
6.7
7987
7457
6.3
7965
Abs.
7999
7970
7.3
7990
7978
7.6
7969
Abs.
Outcrop of coal
D
U
FAULT
N
Abs = Coal absent
Tracks
Boreholes
SCALE:
1 inch = 200 feet

MAP 38

JOIN

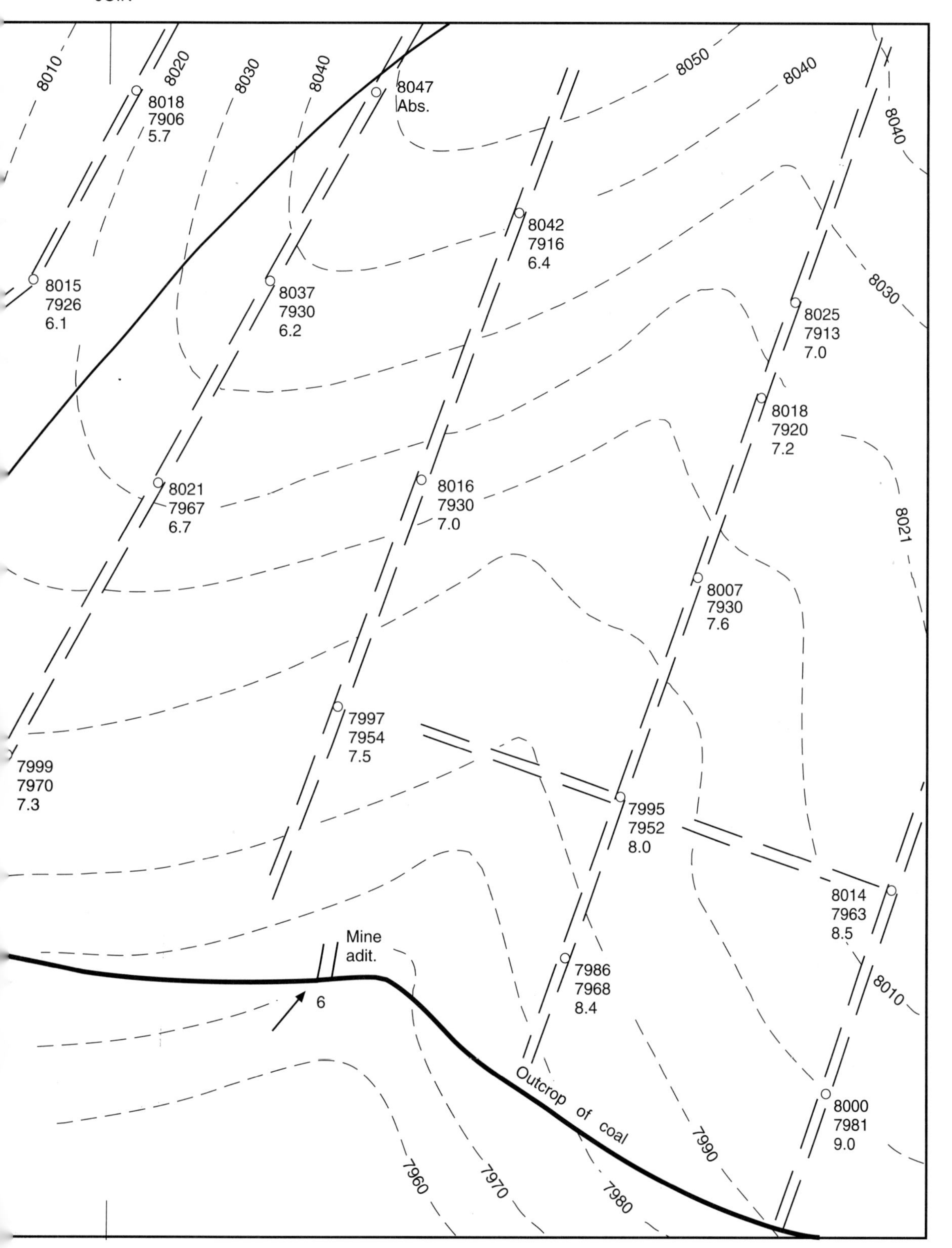

8010
8020
8030
8040
8050
8040
8040
8030
8021
8010
8018
7906
5.7
8047
Abs.
8042
7916
6.4
8015
7926
6.1
8037
7930
6.2
8025
7913
7.0
8018
7920
7.2
8021
7967
6.7
8016
7930
7.0
8007
7930
7.6
7997
7954
7.5
7999
7970
7.3
7995
7952
8.0
8014
7963
8.5
Mine
adit.
6
7986
7968
8.4
8000
7981
9.0
Outcrop of coal
7960
7970
7980
7990

MAP 38 (pages 102–3). A good coal-seam outcrops in a mountainous area of the Sierra Blanca, New Mexico. Shallow boreholes have been sunk at points along lines cut through dense pine forest to add to the meagre information derived from very old mine workings. It is proposed to mine the coal by opencast methods (strip mining). The data are here presented in simplified form: for each borehole three figures are given, namely height of ground above sea level, height of coal above sea level and thickness of coal-seam. All measurements are in feet. As in all consulting problems, rather more data would have been desirable.

Construct a structure contour map for the coal-seam using a 10 ft contour interval. Insert the 50 ft and 100 ft overburden isopachytes on the map. (These will pass through all points where the ground contours and the structure contours for the coal intersect and differ in height by 50 ft and 100 ft. HINT: Naturally, the overburden isopachytes will be approximately parallel to the outcrop of the coal-seam since in this quite small area of the map the topography is not complex. Draw a map to show the isopachytes for the coal-seam thickness using a 'contour' interval of 0.5 ft (6.0, 6.5, 7.0, etc.). Note that 8.2 ft of coal are exposed in the mouth of the old coal adit. You may find it less confusing to draw this map on a tracing paper overlay. Note that the two pages of this map overlap at 'Join'. Either side can be folded and secured with a paper clip so that the join-lines coincide.

MAP 39 The outcrops of the Middle Coal Measures (white) and the Lower Coal Measures (stippled) are shown. The heavy broken lines are the structure contours drawn on the base of the Lower Coal Measures (and have been deduced from borehole data). Construct structure contours for the Lower Coal Measures/Middle Coal Measures junction using outcrop information. (Note that the base of the Lower Coal Measures is folded, therefore the Lower/Middle Coal Measures boundary will also be folded about the same axial planes. Insert these on the map first.) From intersections of the two sets of structure contours deduce the thickness of the Lower Coal Measures at as many points as possible. Draw isopachytes at 50 m intervals for the Lower Coal Measures. (Note that neither structure contours nor isopachytes can be drawn with a ruler in this example.) Drawing isopachytes can be done very conveniently on an overlay of tracing paper. Alternatively, different colours may be used for fold axial traces, structure contours and isopachytes. Draw a section along a north-west–south-east line using a vertical scale of 1 cm = 200 m (a vertical exaggeration of ×2.5).

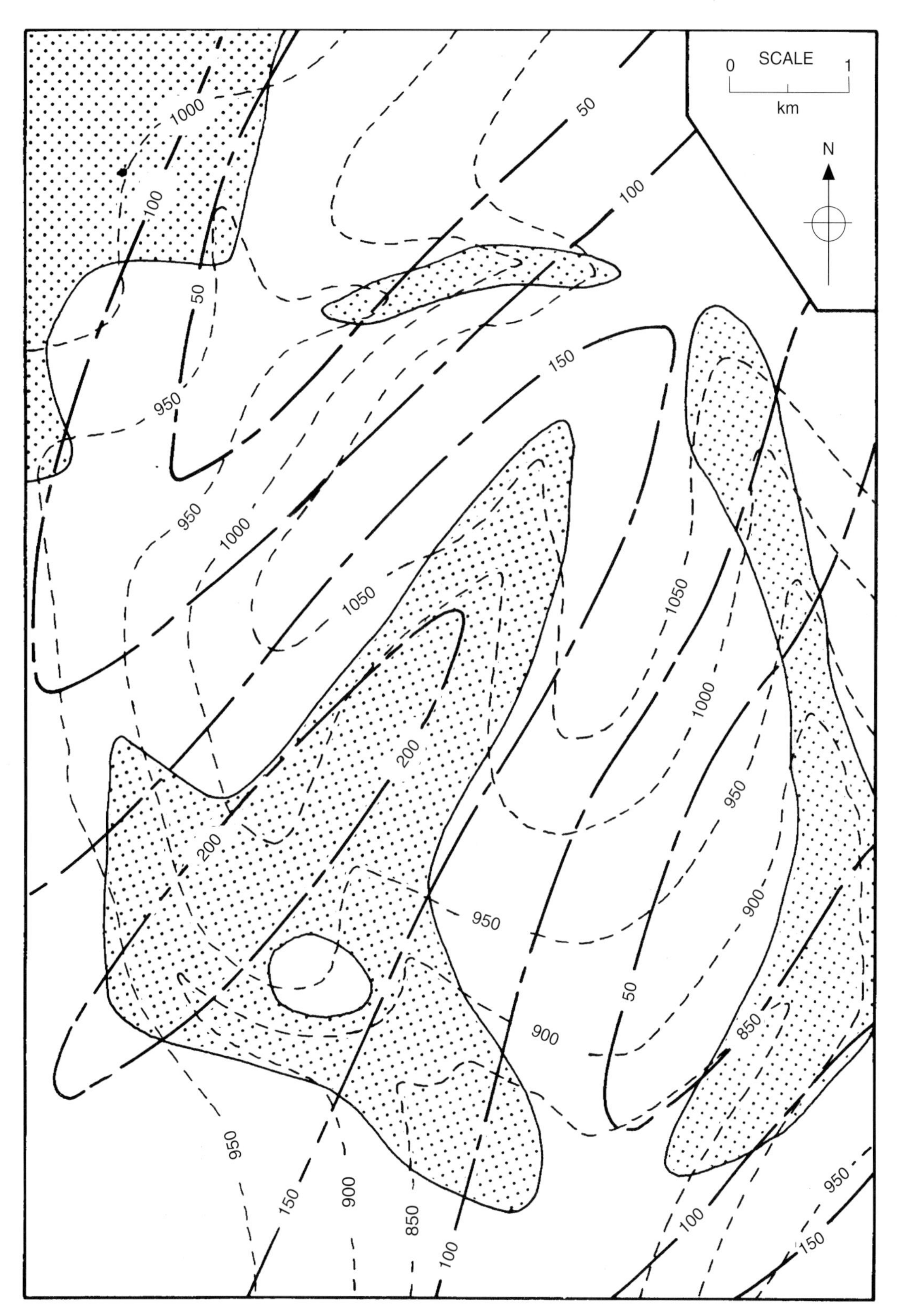

SCALE
0
1
km
N
1000
100
50
950
950
1000
1050
50
100
150
1050
200
200
1000
950
950
900
900
850
50
950
150
900
850
100
950
100
150

Map 39

11

COMPLEX STRUCTURES

NAPPES

Following the discussion in Chapter 4 in which overfolds were described, i.e. structures in which beds are overturned beyond the vertical, we next consider nappe structures. A nappe arises from a very large overfold in which the strata are nearly horizontal over wide areas. The fold structure is referred to as a recumbent fold and has been 'pushed over' so far that both limbs have low angles of dip, and are approximately parallel, although in the case of one limb the beds are actually upside down, i.e. the succession is inverted. Only at the 'nose' of the structure where the strata are folded back on themselves will steep dips be encountered (Fig. 43).

Of course, minor folding is usually superimposed on a major fold such as this so that locally steep dips may be seen. This folding may be termed parasitic. The nature of the folding is related to its position on the overfold (Fig. 43(b)). This minor folding, frequently seen in the field, is of such a scale that it would be revealed only on large-scale maps.

The axial plane of a recumbent overfold is nearly horizontal, but may be curved as in the above figure. Not infrequently the intense lateral stresses producing recumbent folds also

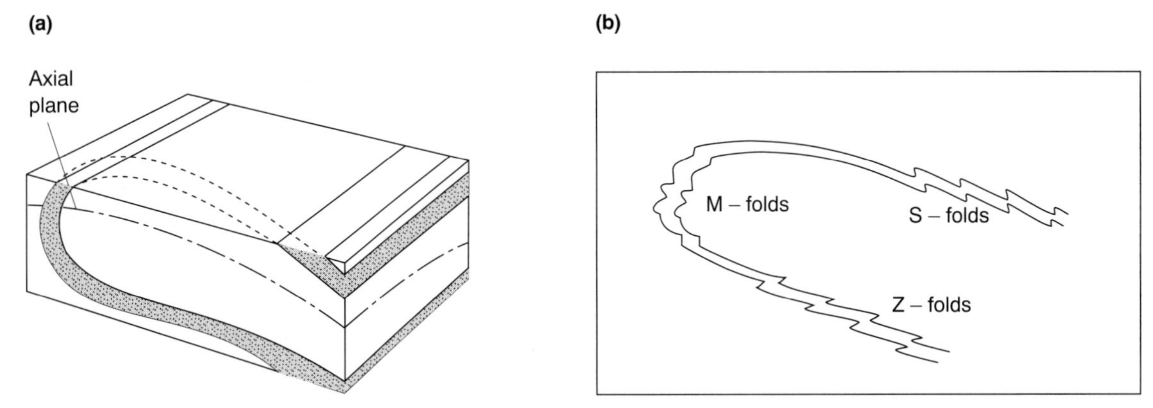

FIGURE 43 (a) Idealized block diagram of a recumbent overfold. (b) The relationship of minor folds to their position on the overfold: M folds occur in the hinge region, S and Z folds are found on its limbs.

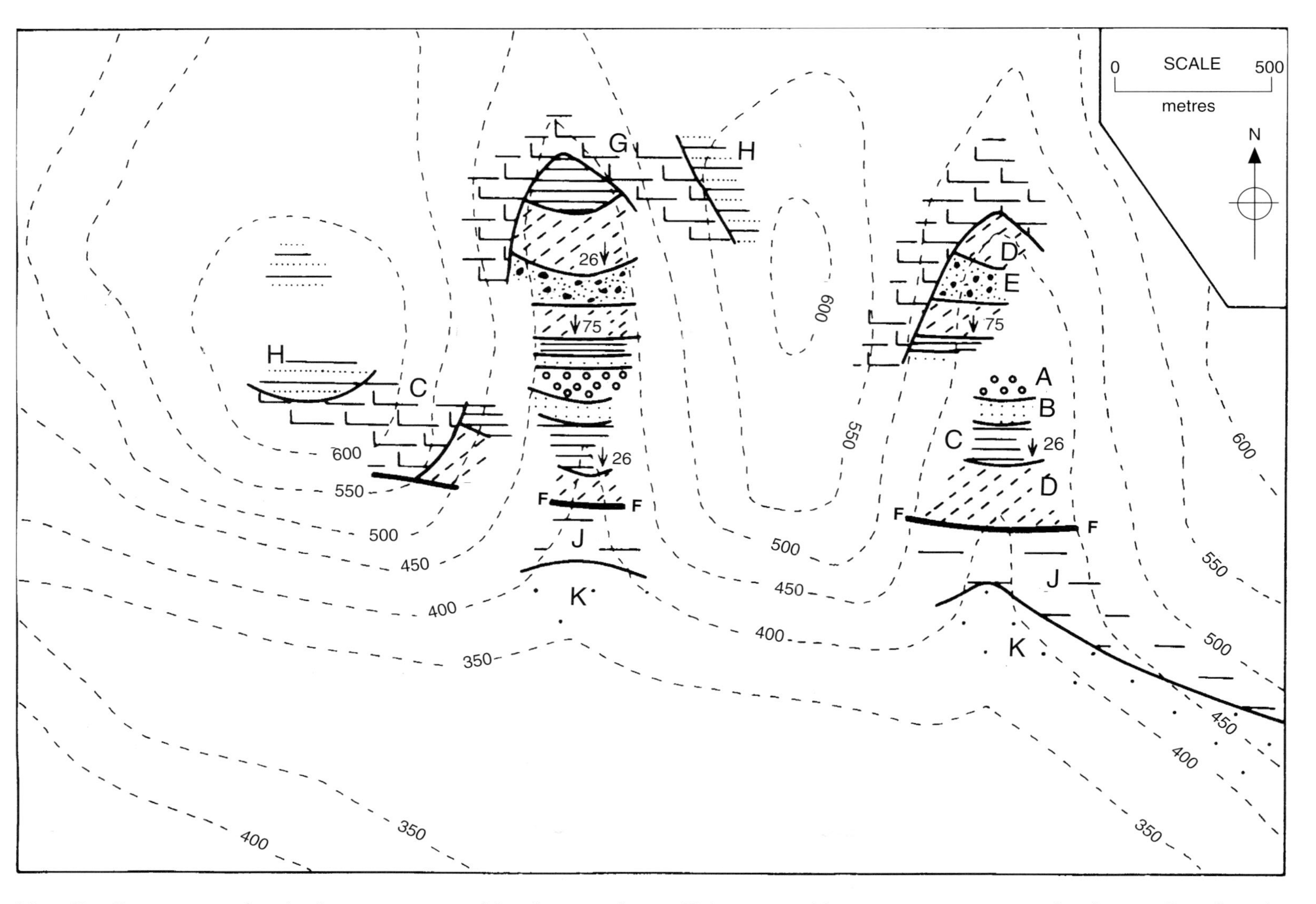

Map 40 Outcrops are few in the area portrayed by the map, but sufficient to enable structure contours to be drawn. Complete the outcrops over the whole map and draw a section along a north–south line. Briefly summarize the geological history.

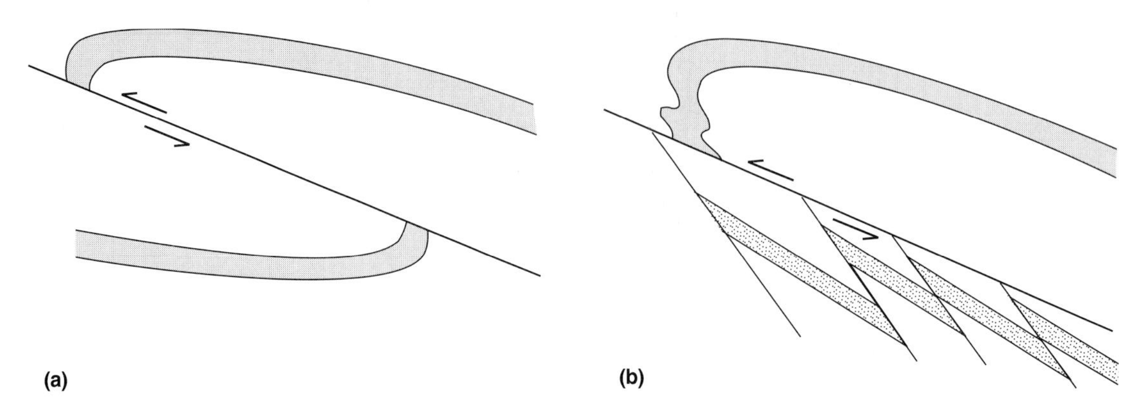

FIGURE 44 (a) Section of a recumbent fold which has been thrust (a nappe) and (b) section showing imbricate structures which commonly occur beneath such a thrust plane.

cause rupture of the strata and produce a 'low angle' reversed fault (making a low angle with the horizontal). The thrust recumbent fold is a nappe structure (Fig. 44(a)).

THRUST FAULTS

A thrust is a low angled fault plane along which movement has taken place, the strata above the thrust plane having been carried often for great distances in a near-horizontal direction, by intense earth movements, over the strata beneath. The thrust strata may have been displaced for a distance of many kilometres and may, for example as in the case of the Moinian rocks of Assynt (Sutherland) above the Moine Thrust, be quite different from any of the other rocks in the same district. On the other hand, the strata above the thrust may be similar to those beneath the thrust, but it should be noted that frequently older rocks may be thrust over younger ones. Thus it is possible to find Precambrian rocks overlying (due to thrusting) Cambrian rocks.

The strata above a thrust plane may be approximately parallel to it but, on the other hand, their dips may be unrelated to the inclination of the thrust plane, a whole block of rocks having been moved en masse. Frequently, just above a thrust plane, dips are locally affected by the movement along the thrust, beds being overturned due to the effects of drag over the rocks beneath the thrust.

The strata beneath a thrust plane may be very greatly affected by the forces associated with the thrusting. The effect on these beds is to produce imbricate structures. These comprise many parallel or near-parallel faults, sometimes of high angle of dip, although they are reversed faults which divide the unthrust area or foreland into 'slices' (Fig. 44(b)).

AXIAL PLANE CLEAVAGE

When rocks that are bedded (possessing what may be termed primary structures) are subjected to pressures – or stresses – they may, as we have discussed, become fractured by faults or they may become crumpled into fold structures. The stress applied to a rock may also cause it to be deformed and new structures are formed such as cleavage and schistosity. Cleavage is developed as a result of shortening of the rock in a direction perpendicular to the cleavage planes with a stretching or extension of the rock in the plane of the cleavage.

Cleavage is often found in rocks that are folded. The greatest shortening of strata due to folding is perpendicular to the fold axial planes and it therefore follows that cleavage planes are parallel to the fold axial planes; hence the cleavage is called axial plane cleavage. (This will not be true for a sequence of beds of different competency nor, of course, in the case of complex refolded folds.)

Cleavage dips, indicated by a symbol such as ↘, should not be confused with dips of bedding planes – to which there may not seem at first sight to be an obvious relationship. This information must not, of course, be dismissed as something clouding the issue: if the cleavage is axial plane cleavage (the most frequently found) there will be a consistent relationship between the cleavage direction and the main structural features (folds). The cleavage will be parallel to the axial planes of the folds.

Maps of areas of more complex structure and with a degree of metamorphism often include information on cleavage. Note that cleavage dips are given on Map 43 and they provide a useful clue in solving the structural problems.

The relationship between the cleavage dip and the dip of the bedding reveals, in overfolds, in which limb the beds are the right way up and in which limb the beds are inverted. Cleavage dip steeper than bedding = right way up; bedding dip steeper than cleavage dip = inverted (Fig. 45). This 'rule' is necessarily true when there has been only one period of folding.

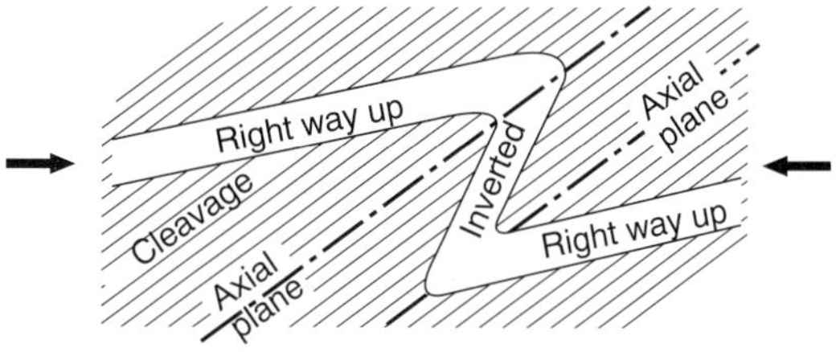

FIGURE 45 Section showing cleavage/bedding relationships in overfolds with axial plane cleavage.

DESCRIPTION OF A GEOLOGICAL MAP

All sources of information available should be used and coordinated: the information deducible from the map itself, the column of strata usually provided and sections showing the geological structures (the latter if not given on the map, drawn by the student). The description of a map should summarize the strata present. The general structural pattern and the trend of the chief structural features should be deduced from outcrop patterns. The broad relationship of topography to geology should be noted. It is usual to summarize the economic geology. Most important, a geological history of the area should be described. This comprises an attempt to date all geological events relatively, the deposition of sediments, breaks in the succession (unconformities), the intrusion of igneous rocks and the development of faults and folds. The evolution of superficial deposits, the drainage pattern and the topography complete the history. In some cases a complete chronology cannot be deduced from map evidence and some events cannot be dated but their possible age and the ambiguity of the evidence must be discussed. Often it is useful to illustrate your answer with a structural sketch map. An example is given of the main deductions which may be made from Map 42.

THE GEOLOGICAL HISTORY OF MAP 42

The earliest bed present is the ashy slate – formed by tuff falling into a sea in which argillaceous sediment was being deposited. A lull in vulcanicity but with continued deposition accounts for the overlying strata, now slate. The age of both formations is in doubt: they are thrust over Carboniferous Basal Conglomerate (which is clearly younger) and they are faulted against volcanic rocks. Though metamorphism itself is no guide to antiquity of

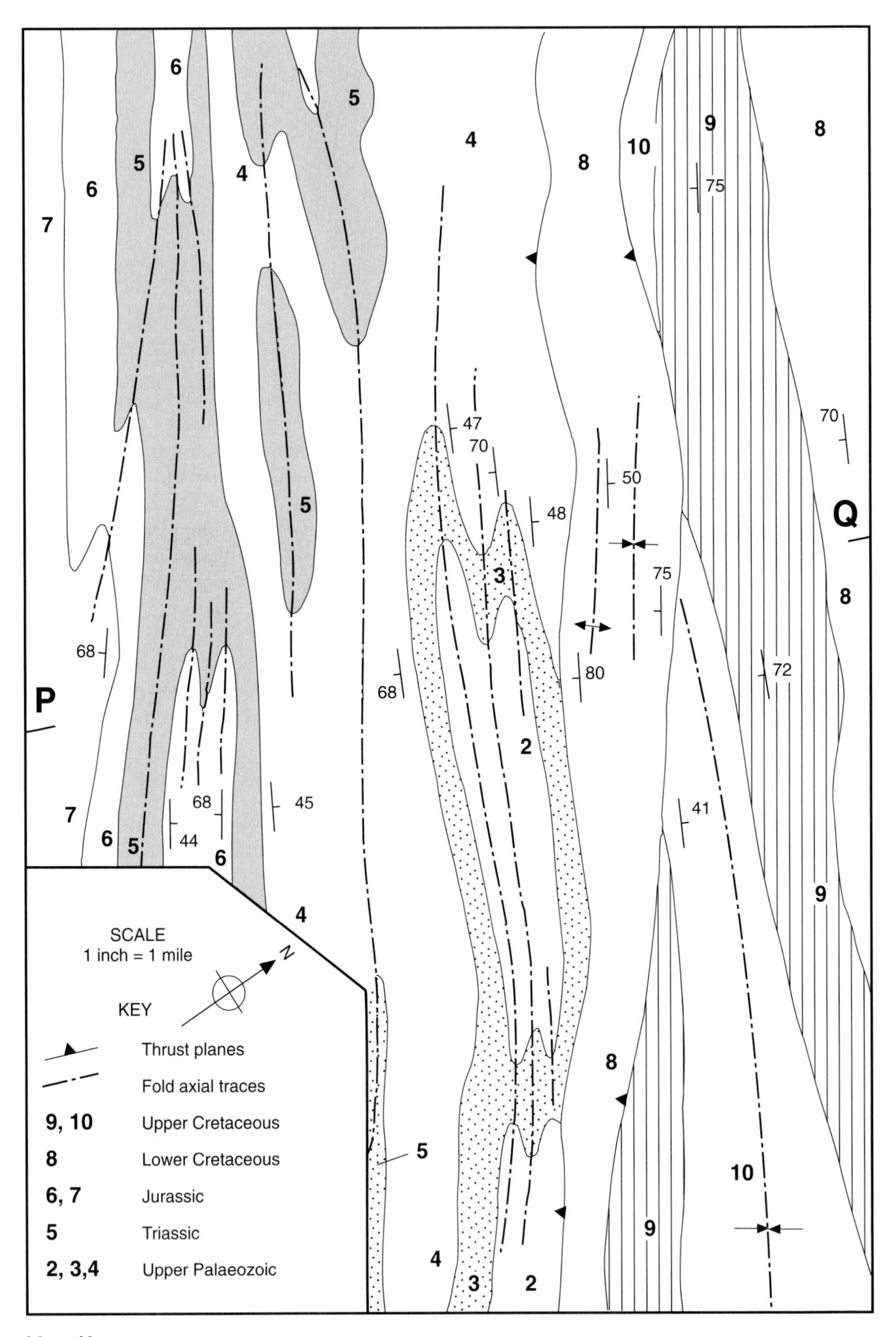
P
Q
SCALE
1 inch = 1 mile
N
KEY
Thrust planes
Fold axial traces
9, 10 Upper Cretaceous
8 Lower Cretaceous
6, 7 Jurassic
5 Triassic
2, 3,4 Upper Palaeozoic

MAP 41

MAP 41 This map is adapted from a portion of the one-inch to the mile sheet 83 E/10 Adams Lookout, Alberta, Canada. It is a highly folded area of Upper Palaeozoic and Mesozoic strata. Some fold structures persist longitudinally for many miles with a roughly north-west–south-east trend (note the direction of north on this map). Other plunging folds, especially minor folds, are only locally developed. There are steeply dipping thrust faults (= reversed faults). They are strike faults with a south-westerly dip of about 50°. In reality relief is considerable, 1200 ft from valley bottom to peaks, but contours have been omitted from Map 41 for simplicity. Draw a cross-section (true scale) along the line P–Q on the profile provided. Most of the fold axial traces are shown on the map. Indicate any others that may have been omitted. Show which folds are synclinal (mark with S) and which are anticlinal (mark with A).

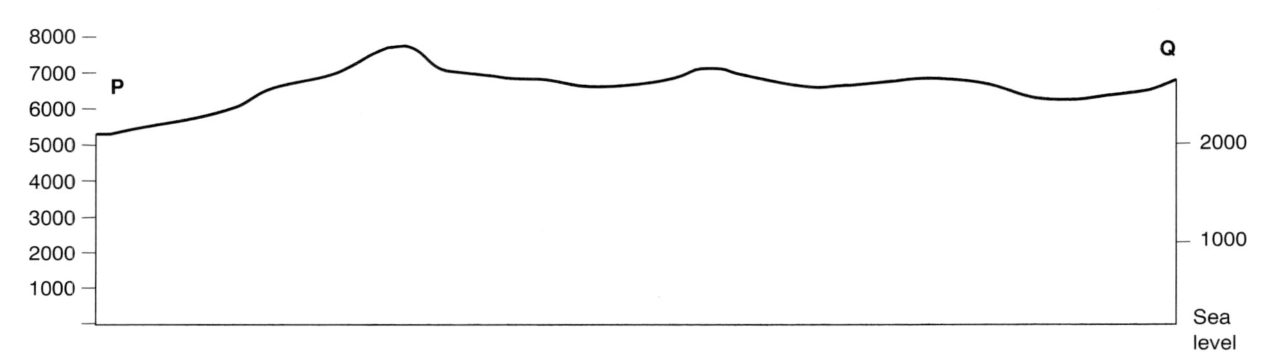

Topographic profile of section P–Q.

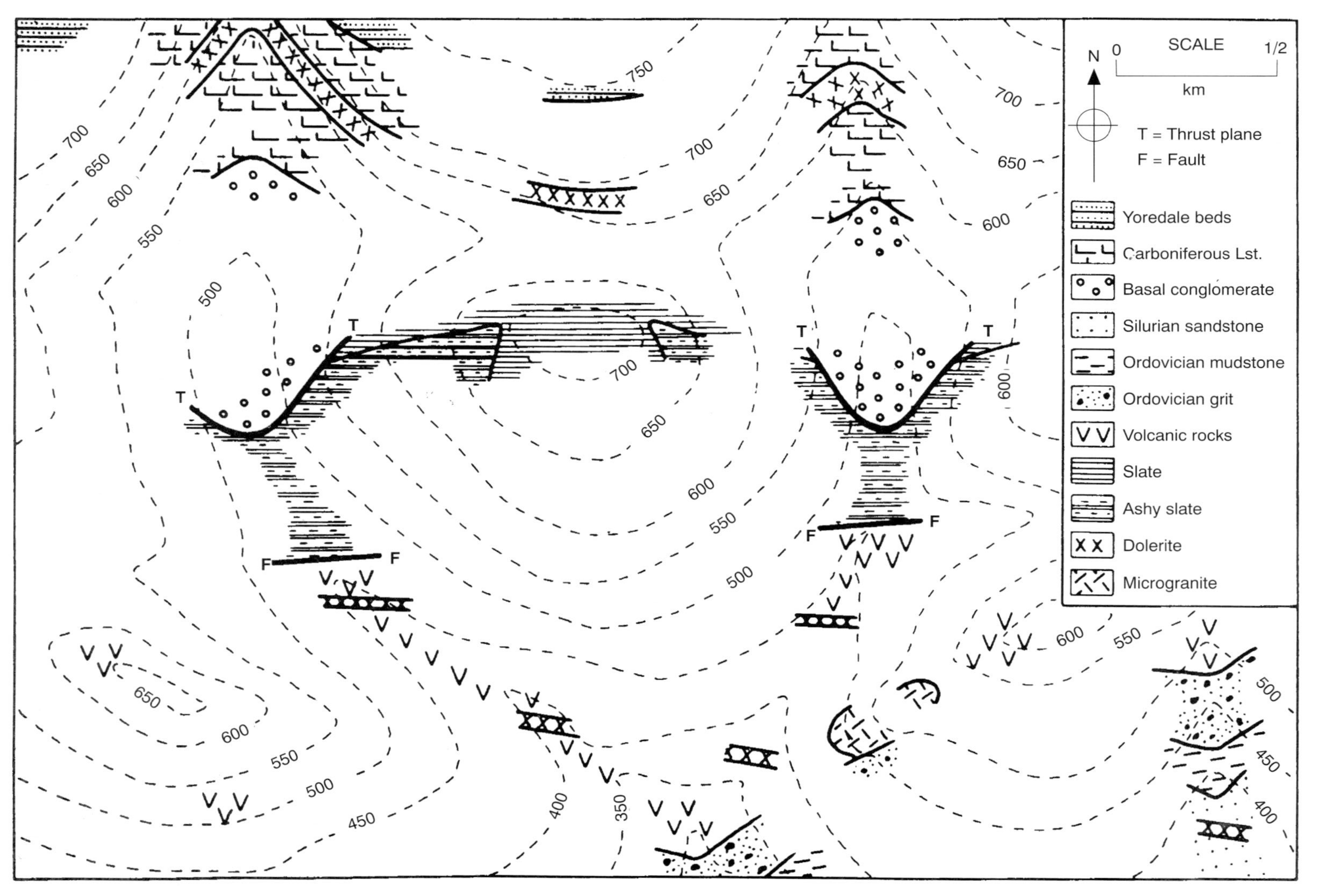

Map 42 This is part of a geological map produced in the field (in northern England). Actual outcrops and exposure of geological boundaries are limited because of vegetation and the presence of superficial deposits of peat, alluvium, etc., not shown on the map. Complete the geological outcrops on the map. Draw a section along a north–south line to illustrate the structures. Write a geological history of the area including notes on the relative ages of the igneous rocks.

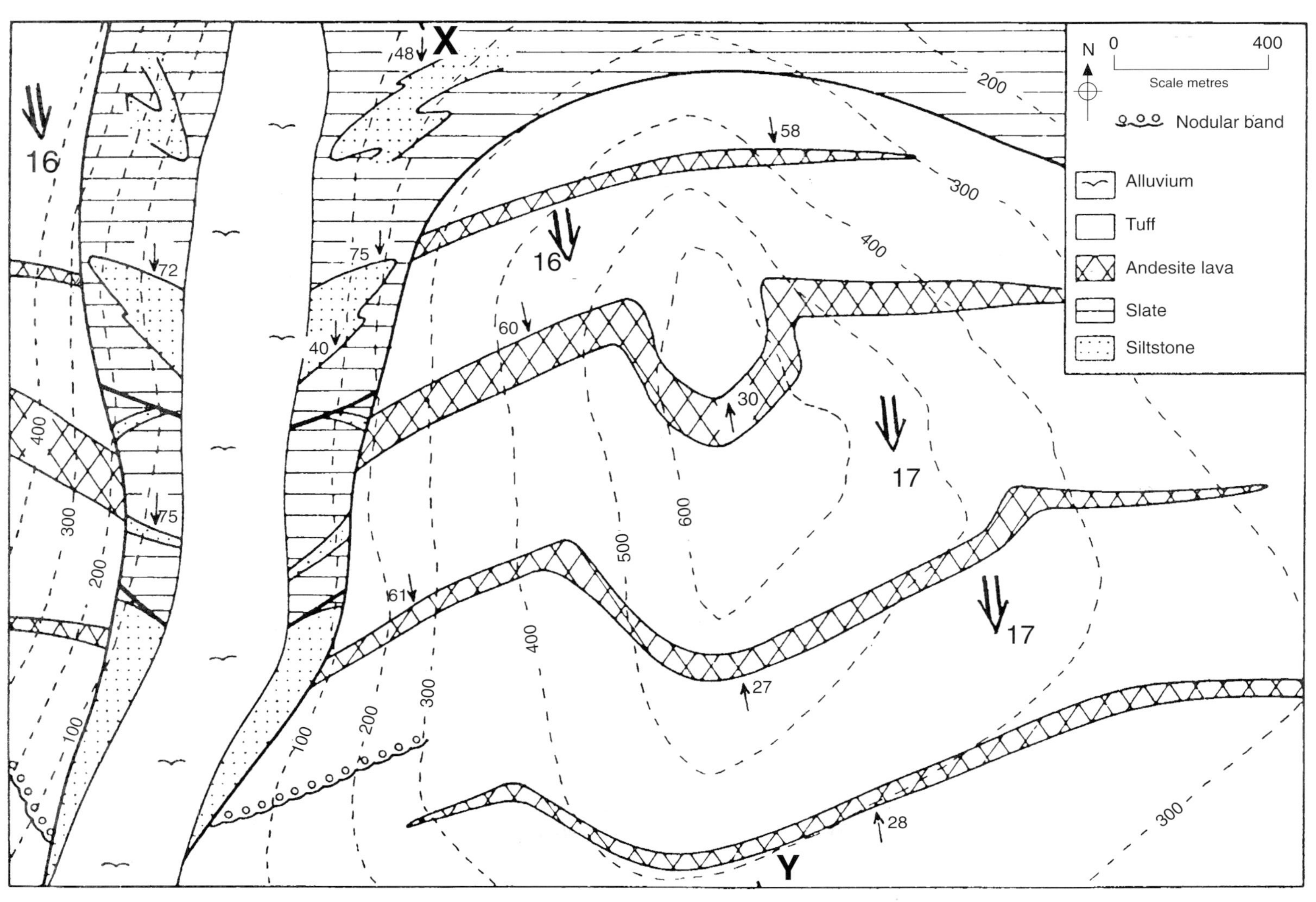

Map 43 Describe the geological structure of the area of the map and draw a section along the line X–Y.

strata, these slates (and ashy slates) may be the oldest strata present, since no other beds have been metamorphosed, and are thus older than the volcanics which are earlier than the Ordovician grit. After the volcanic episode (perhaps prolonged although the thickness of volcanic rocks is not deducible since their structure is unknown) a conformable sequence of sediments was laid down. This comprises Ordovician grit and mudstone and Silurian sandstone. The lowest of these appears to rest partly on the microgranite, which must therefore have been intruded at an earlier date. The age of the east–west fault must be later than the slates and the volcanics which it cuts, but how much later cannot be established.

Dolerite dykes cut volcanic rocks and beds as young as the Silurian sandstone, and hence were intruded after the deposition of these beds. Although not cutting Carboniferous rocks the dykes may be contemporaneous with the post-Carboniferous sill, also of dolerite.

After (probably prolonged) non-deposition (since no Devonian age strata are seen) Carboniferous seas spread into the area depositing initially a basal conglomerate, though its base is not seen and we do not know what it rests on. As the sea became clearer (?deepened) limestone was deposited, followed by the Yoredale beds – cyclic sediments laid down in shallow marine to terrestrial conditions. Two major post-Carboniferous events occurred: the thrusting northwards of older rocks over the Carboniferous, accompanied by overfolding of strata above the thrust plane, and the intrusion of the thick dolerite sill. Neither can be dated precisely. Both may be approximately the same age, referable to the late Carboniferous orogeny (the Variscan) although either could be of much later date. A northerly tilt was imparted to the area since the horizontally deposited Carboniferous strata have a northerly dip.

Mesozoic and Tertiary events are unknown since no strata of this age occur here. However, uplift and subsequent erosion have given rise to the present topography, southwards sloping valleys cutting back into a high east–west escarpment. The topography relates closely to the underlying geology. Superficial deposits are not shown on this map.

EXERCISE ON GEOLOGICAL SURVEY MAPS

1. **Assynt: 1″ Geological Survey map (Special Sheet)** How may the thrust planes be distinguished from the unconformities? Find on the map some examples of imbricate structures.

APPENDIX

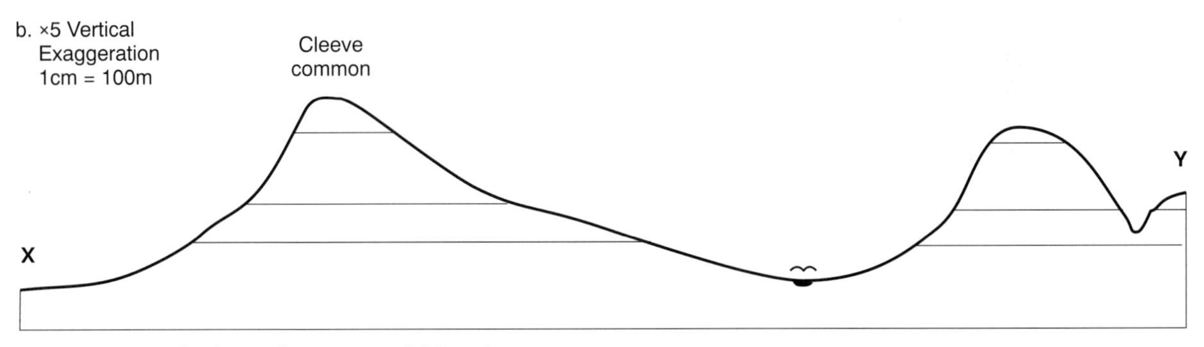

FIGURE 46 Geological sections of Map 2.

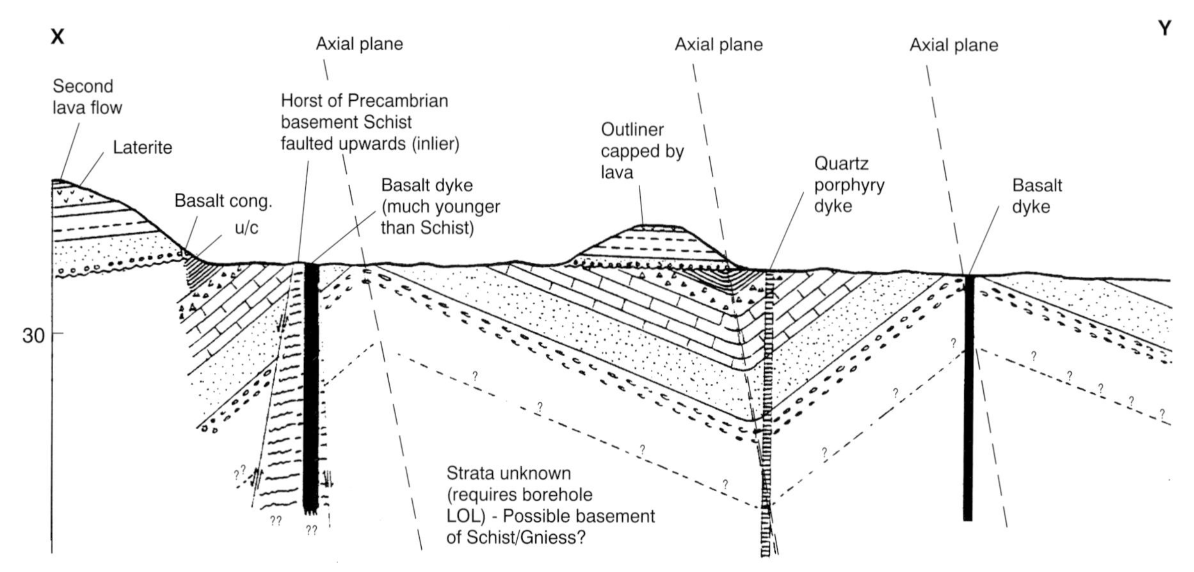

FIGURE 47 Possible solution to section X–Y, Map 23.

TABLE 1 True and apparent dip

The apparent dip of strata, when viewed at an angle to the dip direction, is always less than the true dip. The table below sets out apparent dips at regular intervals. Intermediate values can be interpolated with the graph provided. An example is shown in the table: a true dip of 15 degrees (the left hand column) viewed at 60 degrees to the true dip direction (the top row) appears as a dip of 7.6 degrees (in box).

Dip	Angle of horizontal deviation from true dip direction													
	0	**10**	**20**	**30**	**40**	**50**	**55**	**60**	**65**	**70**	**75**	**80**	**85**	**90**
0	0	0	0	0	0	0	0	0	0	0	0	0	0	0
5.0	5.0	4.9	4.7	4.3	3.8	3.2	2.9	2.5	2.1	1.7	1.3	0.9	0.4	0.0
10.0	10.0	9.9	9.4	8.7	7.7	6.5	5.8	5.0	4.3	3.5	2.6	1.8	0.9	0.0
15.0	15.0	14.8	14.1	13.1	11.6	9.8	8.7	7.6	6.5	5.2	4.0	2.7	1.3	0.0
20.0	20.0	19.7	18.9	17.5	15.6	13.2	11.8	10.3	8.7	7.1	5.4	3.6	1.8	0.0
25.0	25.0	24.7	23.7	22.0	19.7	16.7	15.0	13.1	11.1	9.1	6.9	4.6	2.3	0.0
30.0	30.0	29.6	28.5	26.6	23.9	20.4	18.3	16.1	13.7	11.2	8.5	5.7	2.9	0.0
35.0	35.0	34.6	33.3	31.2	28.2	24.2	21.9	19.3	16.5	13.5	10.3	6.9	3.5	0.0
40.0	40.0	39.6	38.3	36.0	32.7	28.3	25.7	22.8	19.5	16.0	12.3	8.3	4.2	0.0
45.0	45.0	44.6	43.2	40.9	37.5	32.7	29.8	26.6	22.9	18.9	14.5	9.9	5.0	0.0
50.0	50.0	49.6	48.2	45.9	42.4	37.5	34.4	30.8	26.7	22.2	17.1	11.7	5.9	0.0
55.0	55.0	54.6	53.3	51.0	47.6	42.6	39.3	35.5	31.1	26.0	20.3	13.9	7.1	0.0
60.0	60.0	59.6	58.4	56.3	53.0	48.1	44.8	40.9	36.2	30.6	24.1	16.7	8.6	0.0
65.0	65.0	64.7	63.6	61.7	58.7	54.0	50.9	47.0	42.2	36.3	29.0	20.4	10.6	0.0
70.0	70.0	69.7	68.8	67.2	64.6	60.5	57.6	53.9	49.3	43.2	35.4	25.5	13.5	0.0
75.0	75.0	74.8	74.1	72.8	70.7	67.4	65.0	61.8	57.6	51.9	44.0	32.9	18.0	0.0
80.0	80.0	79.8	79.4	78.5	77.0	74.7	72.9	70.6	67.4	62.7	55.7	44.6	26.3	0.0
85.0	85.0	84.9	84.7	84.2	83.5	82.2	81.3	80.1	78.3	75.7	71.3	63.3	44.9	0.0
90.0	90.0	90.0	90.0	90.0	90.0	90.0	90.0	90.0	90.0	90.0	90.0	90.0	90.0	90.0

Table 2 Bed thickness, if vertical thickness of bed (V.T.) = 100 m

Angle of dip (degrees)	True thickness of bed (m)
0	100
10	98.5
20	94.0
30	86.6
40	76.6
50	64.3
60	50.0
70	34.2
80	17.4

Table 3 Outcrop width if true thickness of bed = 100 m

Angle of dip (degrees)	Outcrop width on a horizontal surface
0	infinite
10	575.9
20	292.4
30	200.0
40	155.6
50	130.5
60	115.5
70	106.4
80	101.5
90	100.0

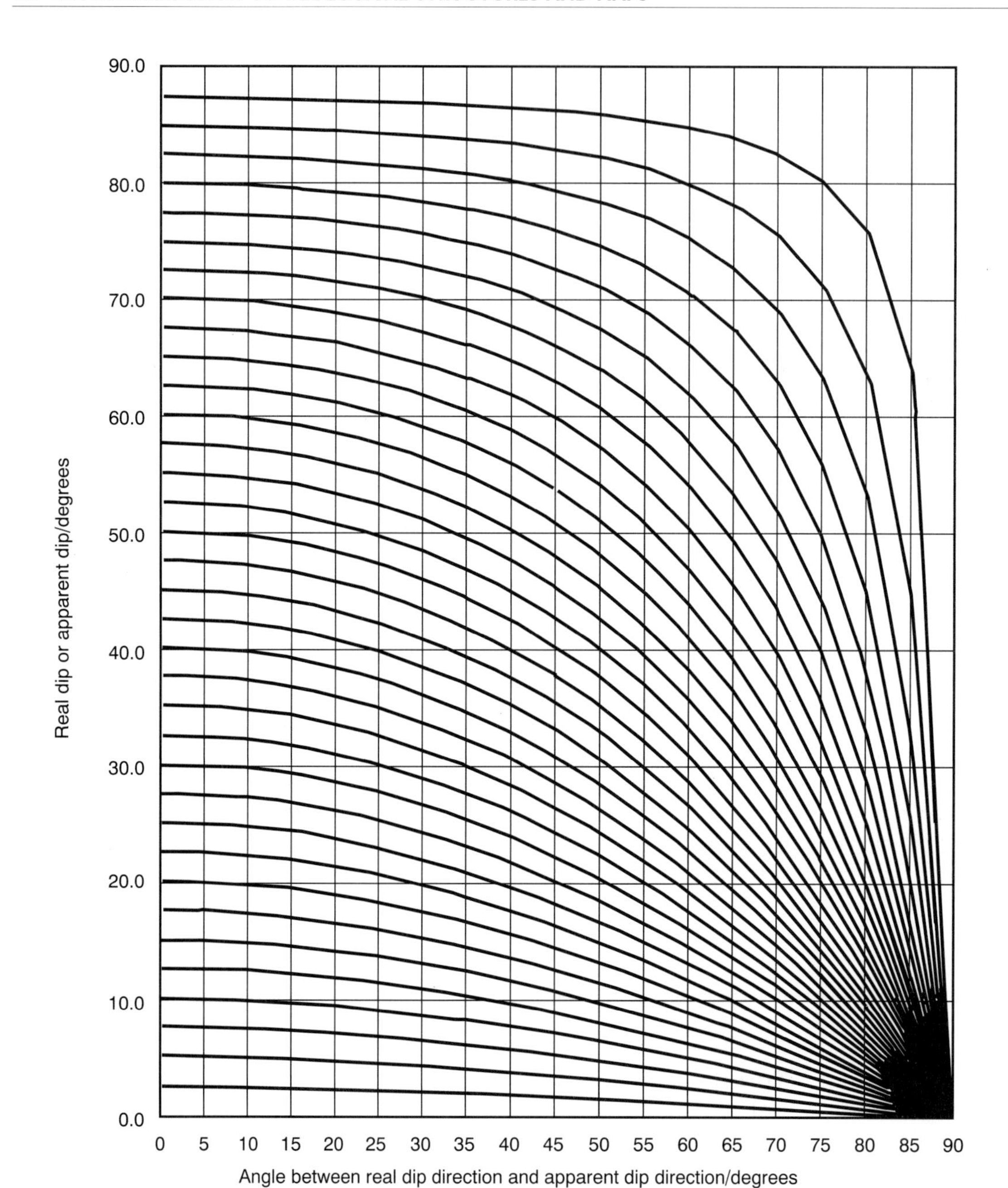

FIGURE 48 Real–apparent dip conversion graph.

NUMERICAL ANSWERS

Map 1	Sandstone 2	100 m
	Mudstone	150 m
	Shale	50 m
	Sandstone 1	150 m
Map 4	1 in 2½; Bearing 174° (6°E. of South)	
Map 5	1 in 5; East	
	B	250 m
	C	100 m
	D	100 m
	E	50 m
Map 6	200 m	
Map 10	A	450 m
	B	200 m
	C	absent
Map 14	500 m	
	200 m	
Map 24	Wrench fault	
	No throw	
	Lateral displacement 3.2 cm = 615 m	
Map 37	Y	400 m
	Z	110 m
	1 in 4 (14°)	

Each of the following eleven diagrams may be cut out and folded to form a box. You then have a three-dimensional model of a geological structure, each side of the block showing a section through the structures outcropping on the surface.

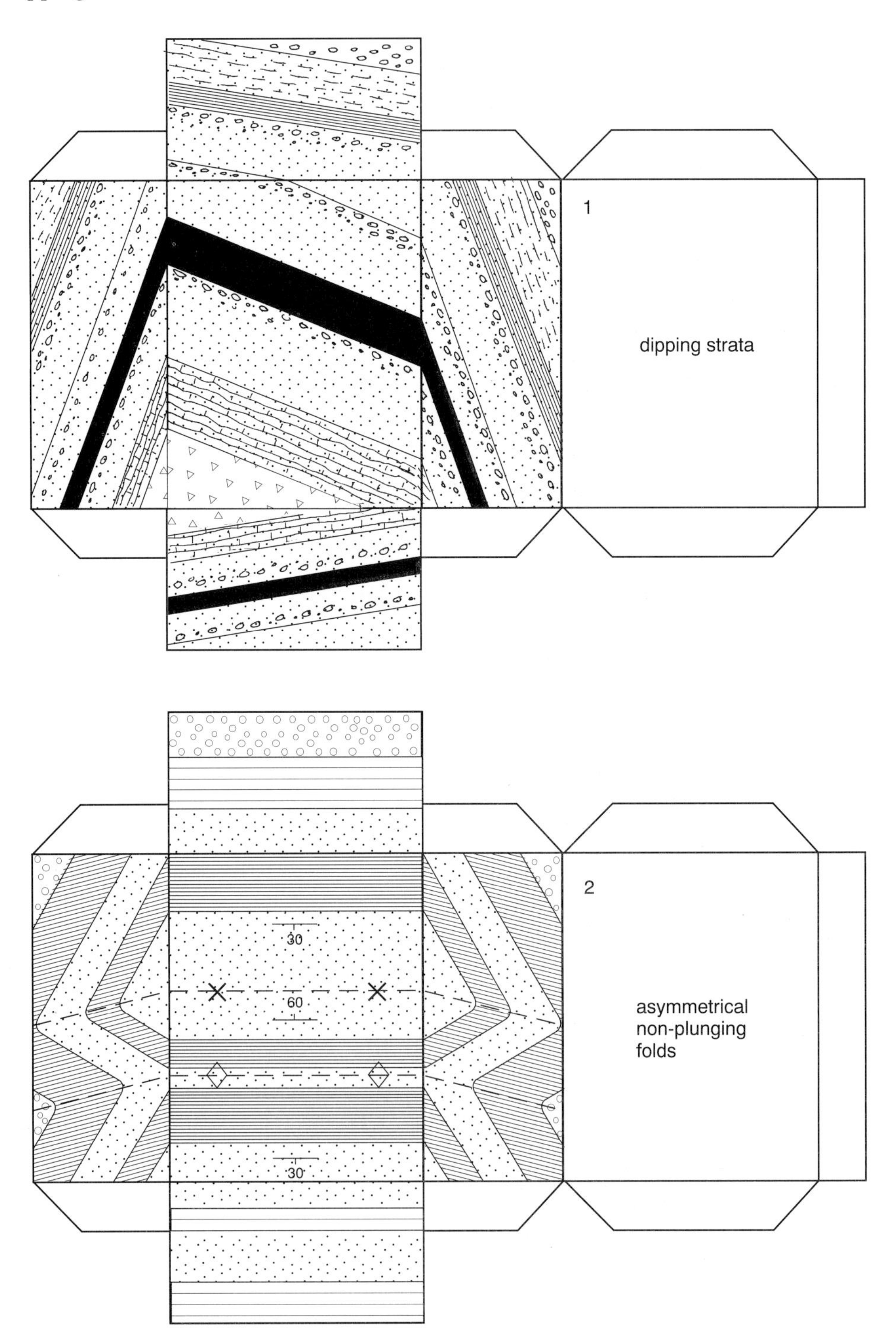

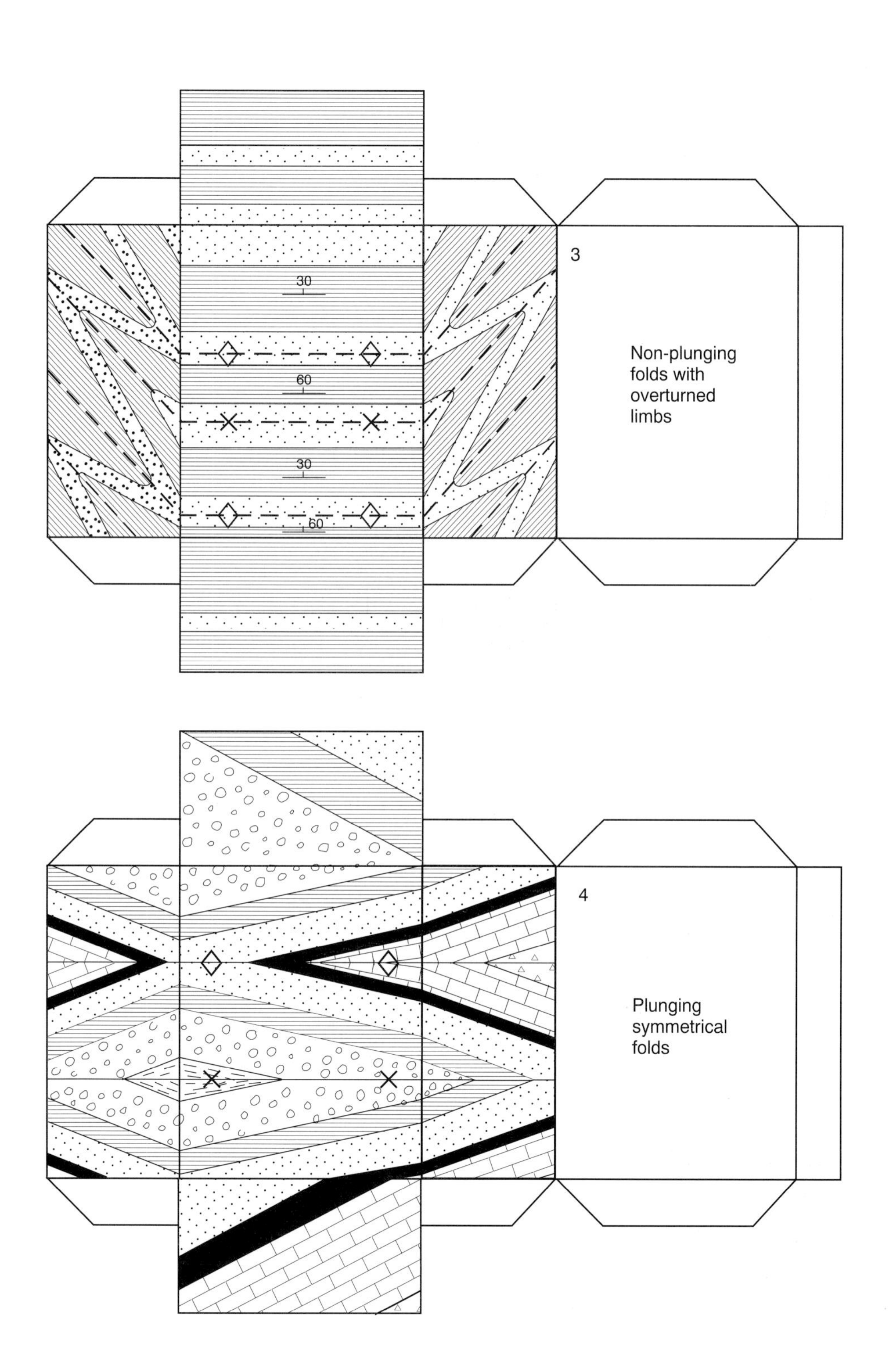
30
60
30
60
3
Non-plunging folds with overturned limbs
4
Plunging symmetrical folds

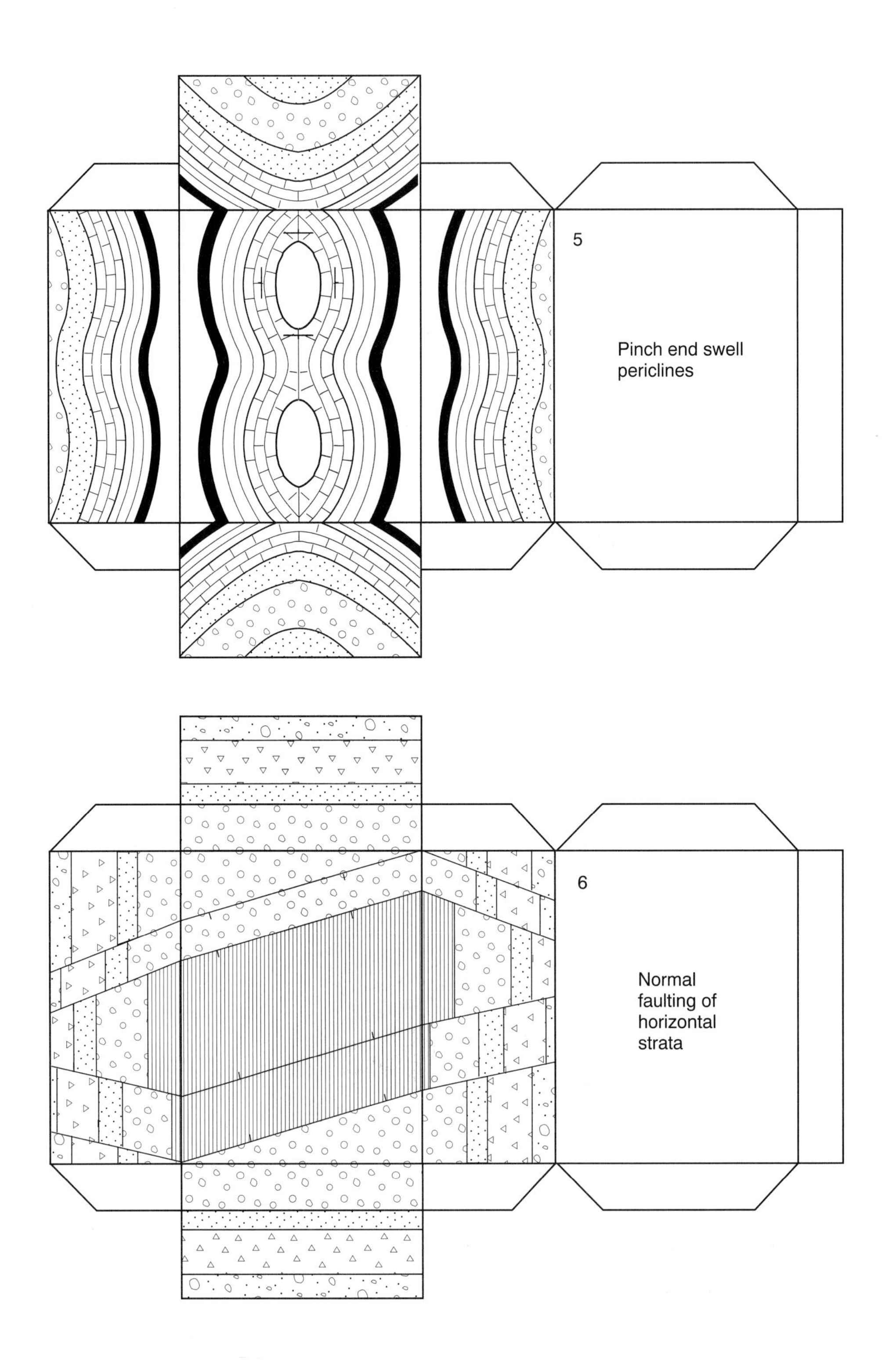
5
Pinch end swell periclines
6
Normal faulting of horizontal strata

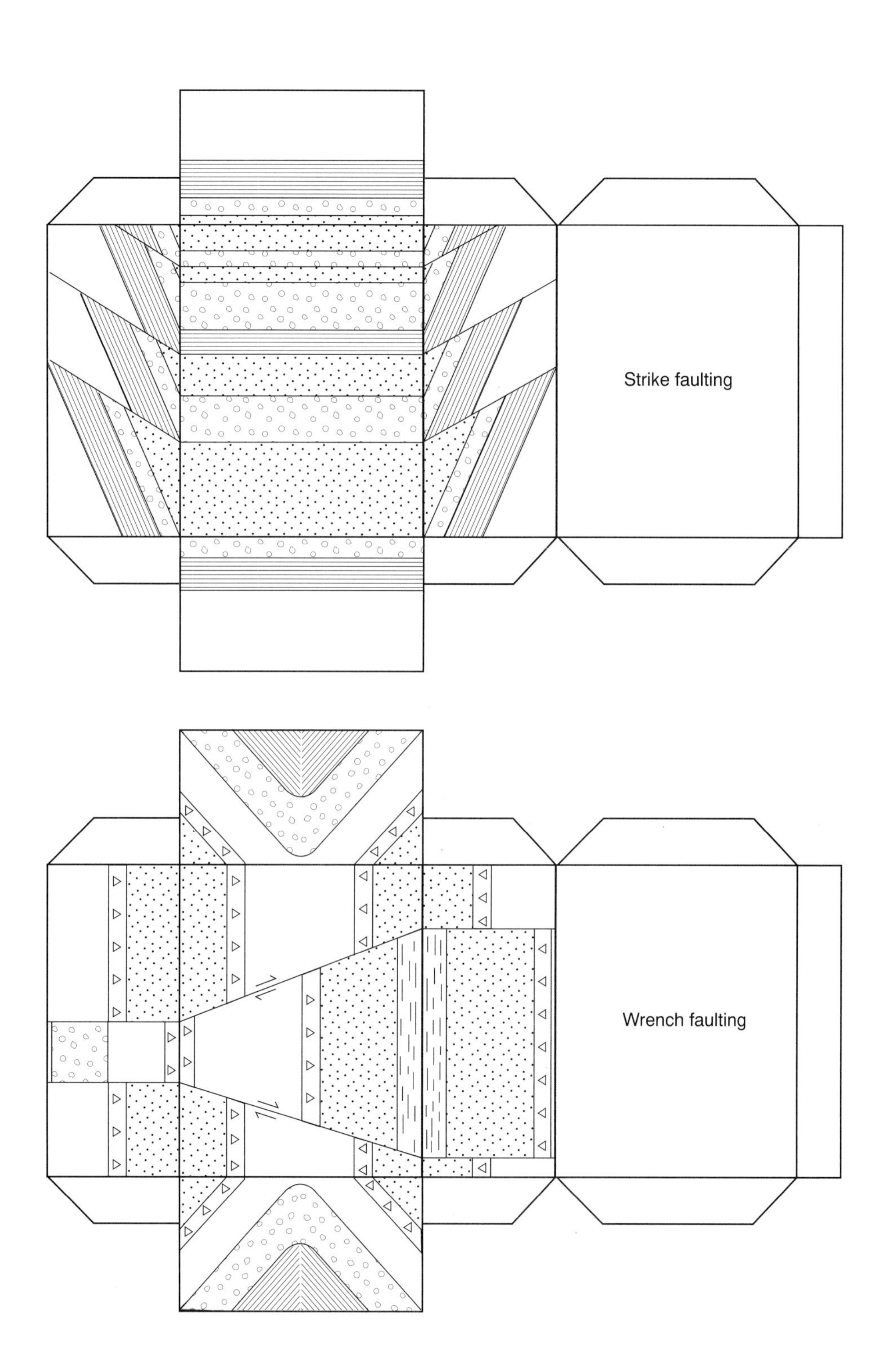
Strike faulting
Wrench faulting

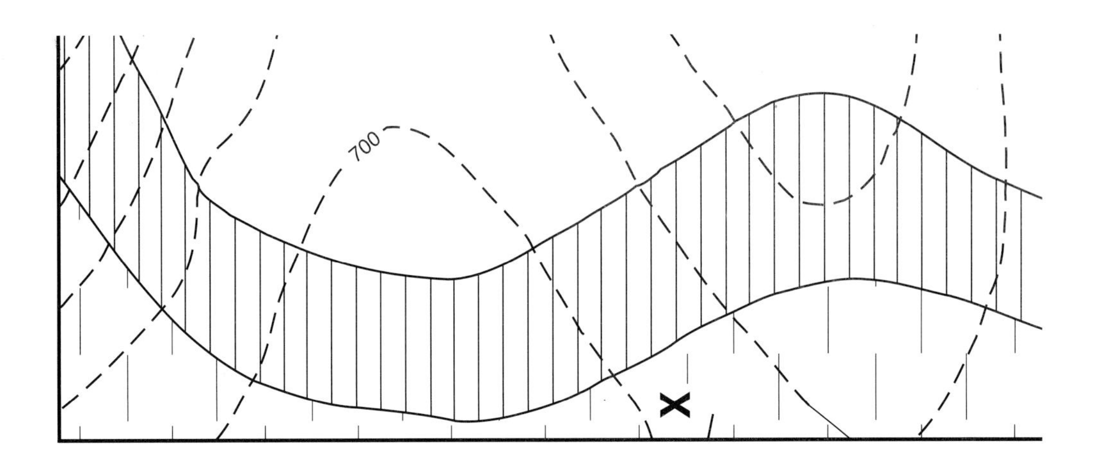
700
X

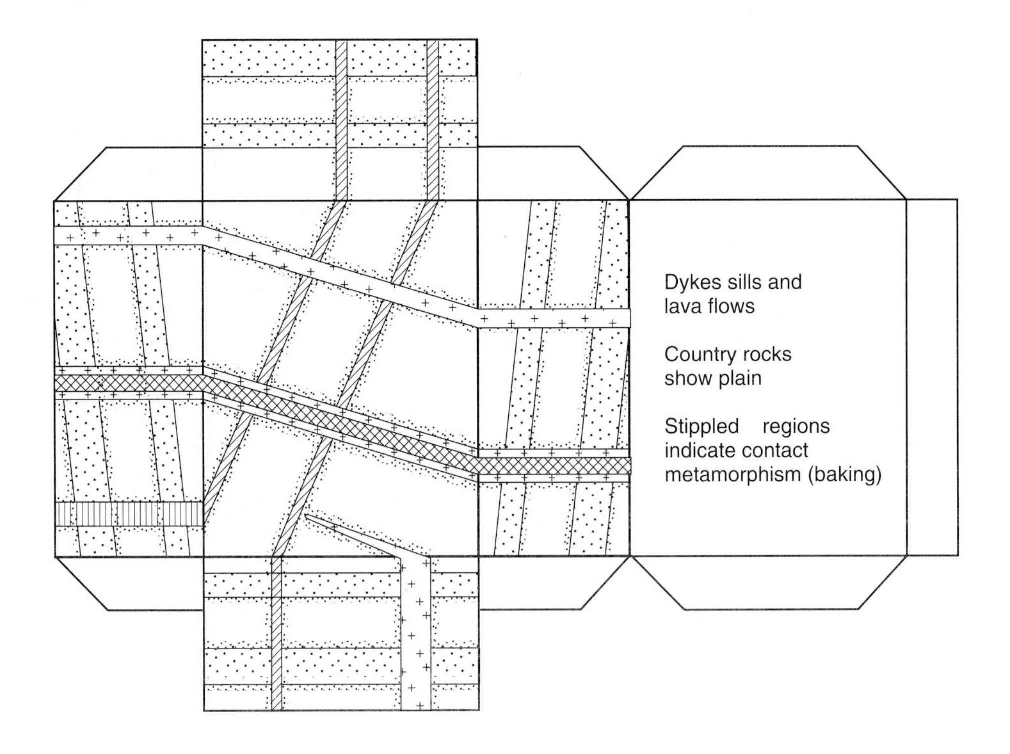
Dykes sills and lava flows
Country rocks show plain
Stippled regions indicate contact metamorphism (baking)

Fig W ptI

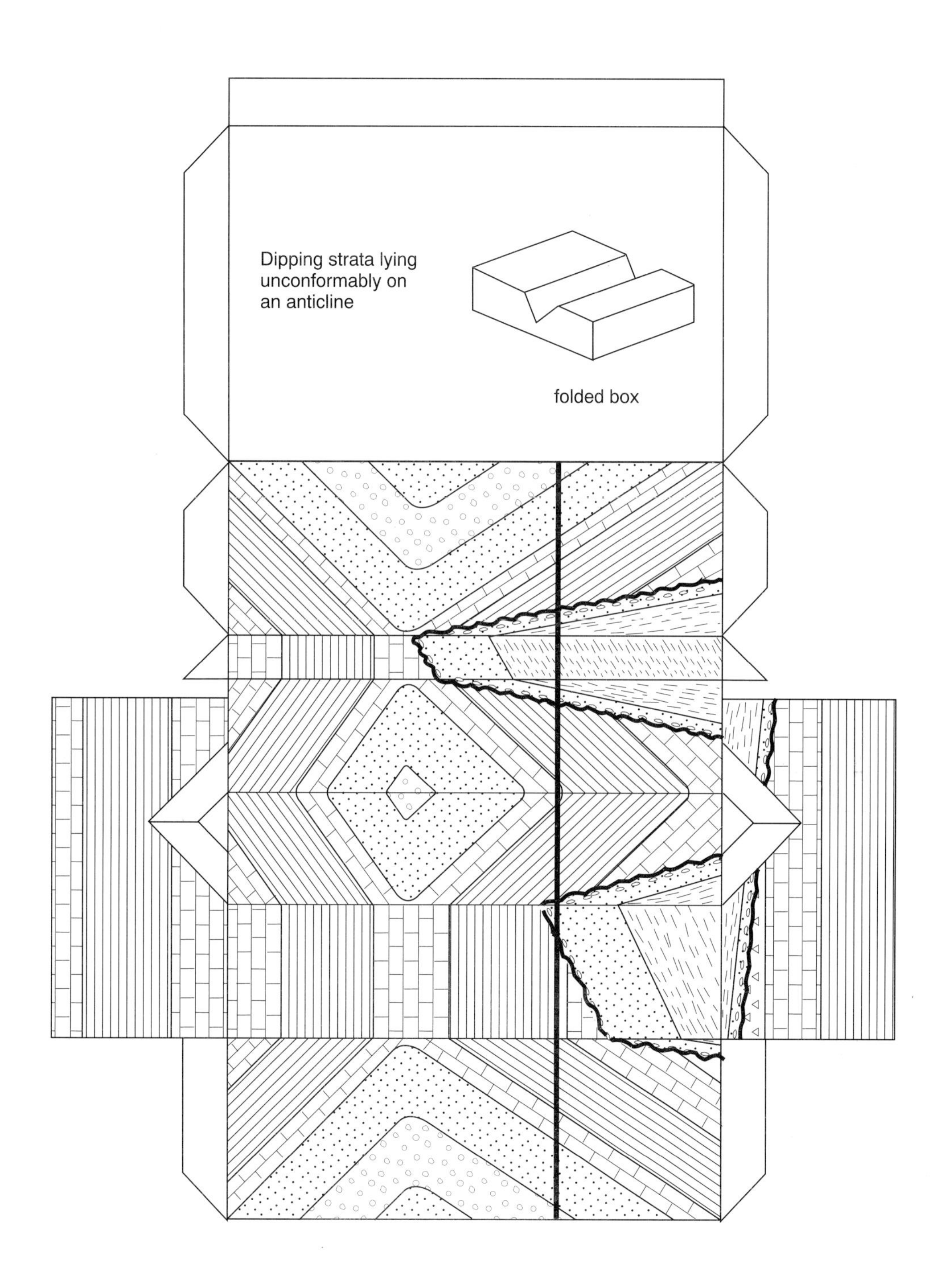
Dipping strata lying
unconformably on
an anticline
folded box

INDEX

Page numbers in **bold** refer to figures and maps.